Asia and the Future of Football

Football is the most popular sport in the world. Globalisation and commercialisation of the game, however, have created new conflicts and challenges. This book explores the role of the Asian Football Confederation (AFC) within the rising significance of football in Asia, drawing on three key theoretical perspectives: globalisation, neo-institutionalism and governance, as well as comprehensive data from interviews and archive material. It explores the organisational structure of AFC, its decision-making processes, relations with other actors, and policies put forward. To understand the specificities AFC has faced in its 60-year history, the broader historical, political, economic, socio-cultural and geographic contexts of football in Asia are taken into account.

Dr Ben Weinberg is an independent scholar and writer based in Cologne, Germany. His areas of interest include political, historical and cultural aspects of sport with a focus on good governance, inclusion and sustainability.

Routledge Research in Sport, Culture and Society

1 **Sport, Masculinities and the Body**
Ian Wellard

2 **India and the Olympics**
Boria Majumdar and Nalin Mehta

3 **Social Capital and Sport Governance in Europe**
Edited by Margaret Groeneveld, Barrie Houlihan and Fabien Ohl

4 **Theology, Ethics and Transcendence in Sports**
Edited by Jim Parry, Mark Nesti and Nick Watson

5 **Women and Exercise**
The Body, Health and Consumerism
Edited by Eileen Kennedy and Pirkko Markula

6 **Race, Ethnicity and Football**
Persisting Debates and Emergent Issues
Edited by Daniel Burdsey

7 **The Organisation and Governance of Top Football Across Europe**
An Institutional Perspective
Edited by Hallgeir Gammelsæter and Benoît Senaux

8 **Sport and Social Mobility**
Crossing Boundaries
Ramón Spaaij

9 **Critical Readings in Bodybuilding**
Edited by Adam Locks and Niall Richardson

10 **The Cultural Politics of Post-9/11 American Sport**
Power, Pedagogy and the Popular
Michael Silk

11 **Ultimate Fighting and Embodiment**
Violence, Gender and Mixed Martial Arts
Dale C. Spencer

12 **The Olympic Games and Cultural Policy**
Beatriz Garcia

13 **The Urban Geography of Boxing**
Race, Class, and Gender in the Ring
Benita Heiskanen

14 **The Social Organization of Sports Medicine**
Critical Socio-Cultural Perspectives
Edited by Dominic Malcolm and Parissa Safai

15 **Host Cities and the Olympics**
An Interactionist Approach
Harry Hiller

16 **Sports Governance, Development and Corporate Responsibility**
Edited by Barbara Segaert, Marc Theeboom, Christiane Timmerman and Bart Vanreusel

17 **Sport and Its Female Fans**
Edited by Kim Toffoletti and Peter Mewett

18 **Sport Policy in Britain**
Barrie Houlihan and Iain Lindsey

19 **Sports and Christianity**
Historical and Contemporary Perspectives
Edited by Nick J. Watson and Andrew Parker

20 **Sports Coaching Research**
Context, Consequences, and Consciousness
Anthony Bush, Michael Silk, David Andrews and Hugh Lauder

21 **Sport Across Asia**
Politics, Cultures, and Identities
Edited by Katrin Bromber, Birgit Krawietz, and Joseph Maguire

22 **Athletes, Sexual Assault, and "Trials by Media"**
Narrative Immunity
Deb Waterhouse-Watson

23 **Youth Sport, Physical Activity and Play**
Policy, Interventions and Participation
Andrew Parker and Don Vinson

24 **The Global Horseracing Industry**
Social, Economic, Environmental and Ethical Perspectives
Phil McManus, Glenn Albrecht, and Raewyn Graham

25 **Sport, Public Broadcasting, and Cultural Citizenship**
Signal Lost?
Edited by Jay Scherer and David Rowe

26 **Sport and Body Politics in Japan**
Wolfram Manzenreiter

27 **The Fantasy Sport Industry**
Games within Games
Andrew C. Billings and Brody J. Ruihley

28 **Sport in Prison**
Exploring the Role of Physical Activity in Penal Practices
Rosie Meek

29 **Sport and Nationalism in China**
Lu Zhouxiang and Fan Hong

30 **Rethinking Drug Use in Sport**
Why the War Will Never be Won
Bob Stewart and Aaron Smith

31 **Sport, Animals, and Society**
Edited by James Gillett and Michelle Gilbert

32 **Sport Development in the United States**
Edited by Peter Smolianov, Dwight Zakus and Joseph Gallo

33 **Youth Olympic Games**
Edited by Dag Vidar Hanstad, Barrie Houlihan and Milena Parent

34 **Safeguarding, Child Protection and Abuse in Sport**
International Perspectives in Research, Policy and Practice
Edited by Melanie Lang and Mike Hartill

35 **Touch in Sports Coaching and Physical Education**
Fear, Risk, and Moral Panic
Edited by Heather Piper

36 **Sport, Racism and Social Media**
Neil Farrington, Lee Hall, Daniel Kilvington, John Price and Amir Saeed

37 **Football and Migration**
Perspectives, Places, Players
Edited by Richard Elliott and John Harris

38 **Health and Elite Sport**
Is High Performance Sport a Healthy Pursuit?
Edited by Joe Baker, Parissa Safai and Jessica Fraser-Thomas

39 **Asian American Athletes in Sport and Society**
Edited by C. Richard King

40 **Pierre Bourdieu and Physical Culture**
Edited by Lisa Hunter, Wayne Smith and Elke Emerald

41 **Reframing Disability?**
Media, (Dis)Empowerment, and Voice in the 2012 Paralympics
Edited by Daniel Jackson, Caroline E. M. Hodges, Mike Molesworth and Richard Scullion

42 **Sport and the Social Significance of Pleasure**
Richard Pringle, Robert E. Rinehart and Jayne Caudwell

43 **A Sociology of Football in a Global Context**
Jamie Cleland

44 **Gambling with the Myth of the American Dream**
Aaron M. Duncan

45 **Inclusion and Exclusion in Competitive Sport**
Socio-Legal and Regulatory Perspectives
Seema Patel

46 **Asia and the Future of Football**
The Role of the Asian Football Confederation
Ben Weinberg

Asia and the Future of Football
The role of the Asian Football Confederation

Ben Weinberg

LONDON AND NEW YORK

First published 2015
by Routledge
2 Park Square, Milton Park, Abingdon, Oxon OX14 4RN

and by Routledge
711 Third Avenue, New York, NY 10017

Routledge is an imprint of the Taylor & Francis Group, an informa business

© 2015 Ben Weinberg

The right of Ben Weinberg to be identified as author of this work has been asserted by him in accordance with sections 77 and 78 of the Copyright, Designs and Patents Act 1988.

All rights reserved. No part of this book may be reprinted or reproduced or utilised in any form or by any electronic, mechanical, or other means, now known or hereafter invented, including photocopying and recording, or in any information storage or retrieval system, without permission in writing from the publishers.

Trademark notice: Product or corporate names may be trademarks or registered trademarks, and are used only for identification and explanation without intent to infringe.

British Library Cataloguing in Publication Data
A catalogue record for this book is available from the British Library

Library of Congress Cataloging in Publication Data
Weinberg, Ben, Dr.
Asia and the future of football: the role of the Asian Football Confederation / Ben Weinberg.
  pages cm. – (Routledge Research in Sport, Culture and Society)
  1. Soccer–Asia. 2. Asian Football Confederation. I. Title.
  GV944.A78 W45 2015
  796.334095–dc23   2014048173

ISBN: 978-1-138-82650-2 (hbk)
ISBN: 978-1-315-73929-8 (ebk)

Typeset in Times New Roman
by Wearset Ltd, Boldon, Tyne and Wear

Printed and bound in the United States of America by Publishers Graphics, LLC on sustainably sourced paper.

Contents

	Acknowledgements	viii
	List of abbreviations	x
1	Asia and the future of football: introduction and contextualisation	1
2	Analysing the Asian Football Confederation: theoretical perspectives and approaches	9
3	Football in Asia: developments and challenges	38
4	AFC's history and origins	79
5	AFC's relations and constellations	85
6	AFC's framework and decision-making	121
7	AFC's policies, strategies and activities	145
8	Understanding the politics of football: discussion and conclusions	171
	Index	187

Acknowledgements

The finalisation of this book required patience and perseverance. It is the compressed version of a doctoral study I conducted over the last few years.[1] Without the assistance and frequent encouragement of many, the realisation of the project would not have been possible. Therefore I would like to take the chance to thank those who provided invaluable advice and support both professionally and personally.

First and foremost, I am indebted to my academic mentors Jürgen Mittag and Fan Hong, both of whom from the beginning of the project showed great interest in my ideas and conceptualisations. They supported my plans and intentions, provided critical and constructive feedback, guided me through challenging phases, took their time for in-depth discussions, and most importantly opened my eyes to the personal development one experiences when undertaking a doctoral degree. The competence of both, and the combination of their expertise in political science, sport history and Asian studies, made them two great supervisors whose enthusiasm for the beautiful game contributed to making the project all the more enjoyable.

Second, I would like to express my gratitude to Karen Petry, Walter Tokarski and all the other wonderful colleagues I had during my time at the German Sport University Cologne. It was Karen who supported my academic development through tuning my skills to strive for high-quality research and teaching, while being a constant source of strength and motivation with regard to my work. Walter's experience, foresight and encouragement meanwhile inspired me throughout all the fruitful exchanges I had with him. Furthermore, I am indebted to Richard Bailey, whose expertise and experience with regard to matters of rhetorical accuracy and scientific coherence have proven to be invaluable in making this book what it is.

For the purpose of collecting data I had the privilege of travelling to Malaysia and Switzerland. Hence I would like to thank AFC and FIFA for enabling access to the archives and ensuring assistance during my stay. I am thankful to the German Academic Exchange Service for providing a grant in order to undertake my visits. In particular I am indebted to Hassan Al Sabah, who made it possible to conduct my research at AFC House. Most notably I would like to thank all the interviewees who took their time to talk to me, give advice and share useful information for the study.

Last but not least I am thankful for how understanding and resilient my family and friends were during the period of the project. Despite me spending hours in front of the computer and bookshelves or being abroad, thinking permanently of what I could and should do, all of them stayed calm and supportive. Without the backing of my parents and my brother – Bernd, Irmi and Joscha Weinberg – as well as my friends – Nils Becke, Peter Beule, Kai van Boxen, Jonas Burgheim, Armin Friedrich, Andreas Hedrich, Svenja Kopyciok, Patrick Manger, Gregor Nentwig, Begona Torres Rodriguez, Aische Westermann and Michael Wiese – I would have never completed this project. Thanks to all of them!

Note

1 The doctoral dissertation was submitted at the German Sport University Cologne in September 2013 and was subsequently defended on 7 February 2014. Univ. Prof. Dr. Jürgen Mittag (German Sport University Cologne) and Prof. Fan Hong (University of Western Australia) supervised the work, while Univ. Prof. Dr. med. Wilhelm Bloch functioned as Chairman of the PhD Committee.

Abbreviations

ACL	AFC Champions League
AFAP	AFC Financial Assistance Programme
AFC	Asian Football Confederation
AFDP	Asian Football Development Project
AFF	ASEAN Football Federation
AIFF	All India Football Federation
ALFC	Asian Ladies Football Confederation
AML	AFC Marketing Limited
ASEAN	Association of Southeast Asian Nations
BINGO	Business international governmental organisation
CAF	Confederation of African Football
CAS	Court of Arbitration for Sport
CFA	Chinese Football Association
CFU	Caribbean Football Union
CONCACAF	Confederation of North, Central America and Caribbean Association Football
CONMEBOL	Confederación SudAmericana de Fútbol
CSCLF	China Soong Ching Ling Foundation
CSL	Chinese Super League
CSR	Corporate social responsibility
DFB	Deutscher Fußball Bund
DRC	Dispute Resolution Chamber (FIFA)
EAFF	East Asian Football Federation
ECA	European Club Association
EPFL	European Professional Football Leagues
EPL	English Premier League
EU	European Union
FA	English Football Association
FAM	Football Association of Malaysia
FAO	Food and Agricultural Organisation of the United Nations
FAS	Football Association of Singapore
FAT	Football Association of Thailand
FFA	Football Federation Australia

Abbreviations xi

FFIRI	Football Federation Islamic Republic of Iran
FFSL	Football Federation of Sri Lanka
FIFA	Fédération Internationale de Football Association
FIFPro	Fédération Internationale des Associations de Footballeurs Professionnels
GAS	General Administration of Sport (China)
GFA	Goa Football Association
GO	Governmental organisation
HK$	Hong Kong Dollar
HKFA	Hong Kong Football Association
IFA	Indian Football Association
IFAB	International Football Association Board
IFRS	International Financial Reporting Standards
IGO	International governmental organisation
INGO	International non-governmental organisation
IOC	International Olympic Committee
IPC	International Paralympic Committee
IPE	International political economy
IR	International relations
ISL	International Sport and Leisure
JAWOC	Japanese World Cup Organising Committee
JFA	Japanese Football Association
KFA	Korea Football Association or Kuwait Football Association
LOC	Local Organising Committee
MOU	Memorandum of understanding
MSL	Malaysia Super League
NAFTA	North American Free Trade Agreement
NFL	National Football League
NGO	Non-governmental organisation
NRA	New regionalism approach
NSC	National Sports Council or National Sports Commission
OFC	Oceania Football Confederation
PFA	Professional Footballers Australia
PFF	Pakistan Football Federation
PIA	Pakistan International Airlines
PPL	Pakistan Premier League
RM	Ringgit Malaysia
ROTA	Reach Out to Asia
SAFA	Singapore Amateur Football Association
SAFF	South Asian Football Federation
SAT	Sports Administration of Thailand
SCAA	South China Athletic Association
TNC	Transnational corporations
TPL	Thai Premier League
UEFA	Union of European Football Associations

xii *Abbreviations*

WADA	World Anti-Doping Association
WAFF	West Asian Football Federation
WAPDA	Water and Power Development Authority
WSG	World Sport Group
WTO	World Trade Organisation

1 Asia and the future of football
Introduction and contextualisation

Football is the most popular game in the world. Its success story relates to how it has adapted to distinctive local interests and contexts around the globe. However, in political terms, globalisation and commercialisation processes have created new constellations, conflicts and challenges for sport in general and football in particular. In fact, the debate on the relation between sport and politics has intensified in recent years. If one inquires of an athlete or club official, then sport as an activity and competitive field of action is most often regarded as apolitical. At the same time, others have expressed their belief in the cultural and political synergies sport is presumed to release. Governments and states use sport as means of public diplomacy or enhancing national identities and collective memories. The French philosopher Robert Redeker (2008: 499–500) has argued, however, that it is 'an illusion to think of sports as [a] controllable' tool since it 'is a total system, a planetary machine that profoundly transforms men and women as well as their relation to the world'. He claims that sport 'has devoured politics [...] while facilitating a theatre of international relations parallel to that of diplomacy'. Due to its supposed dominance over other societal driving forces, he concludes that it 'is the total contemporaneous social myth that has enlisted politics and international relations in its service'.

It is hardly surprising that questions about spheres of influence and the power exercised within and through the organisational capacities of sport have received public attention as well as extensive media coverage. With regard to the beautiful game, eyes have turned to organisations such as FIFA and the Asian Football Confederation (AFC). Following the nomination of Qatar as host country for the World Cup in 2022 and the subsequent competition between Sepp Blatter and Mohamed Bin Hammam for the FIFA Presidency in 2011, ethical and organisational aspects of governing world football have been in the spotlight. It seems the public not only wants to better understand the relation between sport and politics, but also the political processes taking place within the sport sector. Under consideration of the emergence of new powers in the sport policy arena, attention has been drawn to the roles of specific actors as well as structural and cultural characteristics.

In fact, the global awareness of regions like Asia has grown and perceptions of its sport and football culture have changed. This is not only linked to Asia's

2 *Introduction and contextualisation*

rise to economic and political power but in terms of football, for instance to the impact of the World Cup 2002 in South Korea and Japan. Sepp Blatter expressed his belief that Asian football will play an eminent role in the future of world football when he said: 'In Asia you have more than half the world's population and the future of football must be in Asia' (quoted in Asian Football Confederation, 2002: 33). Another example of this perception is AFC's official slogan 'The Future is Asia', which resonates with its major policy objective, namely the aim to facilitate sustainable development and professionalisation of football in Asia (BBC Sport, 2004).

As the dust settled from the successful outcome of the World Cup 2002 in South Korea and Japan, the view of Asian football was on its way to significant reform. Except for North Korea's advancement to the quarter-finals in the World Cup 1966 in England, Asian male football teams had performed to a relatively low level at international football tournaments, and therefore the contribution to spreading and preserving the popularity of football across Asia remained rather marginal. Despite the fact that some Asian players such as Cha Bum-Kun, Park Ji-Sung, Hidetoshi Nakata or Ali Daei made it to the big leagues in Europe, Asian footballers had often been criticised by western coaches for not having the quality to compete on an international scale (Kramer, 2005; Giulianotti and Robertson, 2007: 181; Murray, 1995: 141). This opinion has changed in recent years, taking into consideration the increase of Asian players in leagues like the German Bundesliga where players such as Shinji Kagawa have been pillars of success. This process has led to debates on how Asian football should deal with the migration of players to other regions (Ehrmann and Klappenbach, 2011; Duerden, 2011).

In contrast, female Asian national squads have had a considerable impact on the development and significance of women's football in Asia, with teams such as China having performed successfully during its 'golden age' from 1986 to 1999 (Hong and Mangan, 2003: 52). Japan's Women's World Cup victory in 2011 constitutes a reflection of this development. In terms of legitimisation and commercialisation, however, women's club football in Asia is still trailing behind its male counterpart.

Former AFC General Secretary Paul Mony Samuel (2008) has noted that, in comparison with other regions, football in Asia has developed differently in terms of cultural, social, economic and political aspects. He argued that the size of the continent has been counterproductive to a coherent and systematic development, unlike in Europe and elsewhere. In addition, the dominance of European football through extensive media coverage and marketing strategies has been a further impediment to the growth of Asian football. It thus requires fostering grassroots and youth development, coach education, the emergence of role models, the improvement of structures and relevant frameworks as well dealing with match fixing, bribery and corruption (Manzenreiter and Horne, 2007: 565, 573). When embarking upon the nomination of South Korea and Japan as host countries for the World Cup 2002 and after the subsequent success of that tournament, the AFC therefore formulated an objective to improve all aspects of

Asian football, at both the grassroots and professional levels. Through launching its football development programme 'Vision Asia' in 2002, AFC began implementing its mission in collaboration with a selection of member associations. Further policy initiatives were launched in the form of a financial aid programme and a social responsibility campaign (Weinberg, 2012: 546–548).

From the 1970s onwards and since the beginning of the Presidency of João Havelange in FIFA, the global infrastructure of football has undergone a considerable transformation in terms of shifting from Eurocentric dominance to a conflict of power between traditional and peripheral football nations. This discrepancy of interests found increasing expression during the 1990s, when the traditional pyramid of football government was affected by a web of new relations between miscellaneous private, economic and political actors who were to gain more and more influence and significance. Aspects of commercialisation, broadcasting rights and financial resources developed into central driving forces accompanied by legal and political matters. Federations and governing bodies saw themselves confronted with new opposing players and therefore had to start reconsidering their traditional autonomy while attempting to maintain their political power. In fact, the growing friction between autonomy and regulation has been accelerated by recent debates surrounding accusations of corruption, bribery and manipulation that have not only been spotlighted by the media but by national governments and the European Union. Additionally, representatives of associations and confederations have expressed concerns with these issues and have been looking for ways of intervening in and eroding the traditional political system of world football. Therefore governing bodies have seen themselves confronted with a range of stakeholders that contains potential for conflict between federations, governmental actors, players, clubs, leagues, fans, sponsors or media corporations, while the task of organising competitions and supporting the development of football in terms of education, administration and management, as well as sustaining its cultural matrix, remains central to the vitality of governing bodies (Lee, 2003: 114–116; Giulianotti and Robertson, 2009: 114–116; Henry, 2007: 8; Houlihan, 2003: 31; Groll, 2007: 188–189; Schulze, 2002: 17).

Research specifically into football in Asia continues to be generally limited, although the number of publications that cover it (which are more related to the general topic of football) is relatively high. Most studies concerned with Asian football have concentrated on aspects such as the development of the game itself, or its technical, administrative and business-oriented implications, depending on the academic perspective (e.g. Andreff and Szymanski, 2006; Beech and Chadwick, 2004; Dobson and Goddard, 2001; Hamil *et al.*, 1999). Few studies have been concerned with cultural and societal aspects (e.g. Armstrong and Giulianotti, 1999; Hopf, 1998; Sugden and Tomlinson, 2002; Guttmann, 1994). Even fewer have aimed at offering a political science approach to the topic (e.g. Fanizadeh *et al.*, 2002; Wagg, 1995; Mittag and Nieland, 2007) and those that have explored this field have tended to focus on the European game (e.g. Holt, 2009; Garcia, 2008; Strünck, 2007; Gammelsæter and Senaux, 2011).

4 Introduction and contextualisation

The role of international governance in football against the background of globalisation has been part of Giulianotti's studies (2007). By conducting a multi-dimensional and transnational examination of football with regard to its history, class, gender, ethnicity and nationalism, fans, finance and television, plus the often neglected evolution of aesthetics as a mirror of cultural specificities, Giulianotti produced a capacious account which has been extended through collaboration with Robertson (2009). This resulted in the creation of a theoretical model for the sociological assessment of the development of football. Nonetheless, these works lack a distinctive political science analysis of football in general and Asian football in particular.

Meanwhile, historical accounts of the governing body FIFA have been provided by Eisenberg *et al.* (2004), while the power relations behind FIFA have caught the interest of Sugden and Tomlinson (1998; 1999). In addition, Darby (2002) has worked on FIFA and football in Africa, including the role of the Confederation of African Football (CAF), through delivering insights into political structures, procedures and particularly the power relations in world football. Also Sugden and Tomlinson (1997) as well as Darby *et al.* (1998) have dealt with the relation between FIFA and UEFA and their struggle for control. However, apart from one chapter in Sugden and Tomlinson's (1998) book on the global governing body FIFA and a couple of passages in their follow-up publication *Badfellas* (2003), AFC as a topic remains underexposed in research. Few publications have specific regard to the development of football in Asia. Dimeo and Mills (2001) have edited a book on football in southeast Asia, while the role of football in the Far East has been in the focus of the works of Horne and Manzenreiter (Horne and Manzenreiter 2002; Manzenreiter and Horne 2004). In addition, Majumdar and Bandyopadhyay (2006a; 2006b) have authored two books on football in India.

This book is concerned with the question of which role AFC plays in the infrastructure of Asian and global football. The objective of this examination inevitably connects to the question of how to analyse AFC, and whether the approach employed provides an analytical benefit for sport policy research focusing on actors, structures, interactions and constellations within and outside of sport governing bodies. Based on the tripartite analytical approach examining polity, politics and policy (Lösche, 2010), this analysis takes a look at the organisational structure of AFC and its institutional environment, the relevant decision-making processes taking place around and within AFC, and the programmes and specific policies formulated and implemented by the confederation. This includes an assessment of key actors on all analytical levels, an identification of policy aspirations, priorities and value systems, and an examination of available resources and dependencies in order to evaluate power relations, discourses, interactions and the environmental setting. Furthermore, these aspects and the strategies and instruments employed by AFC in the sport policy delivery process are assessed alongside the broader historical, political, economic, socio-cultural and geographic contexts and structures that shape the policies in order to portray possible path dependencies and development trends.

Introduction and contextualisation 5

While focusing on the determinants, strategies, processes, contents and instruments in order to explore, describe and explain the role of AFC in the infrastructure of football politics, this study cannot provide a substitute to an exhaustive evaluation of actual policy implementation, its outcomes and its effectiveness. It is necessary to build upon a selection of theoretical and conceptual frameworks that are put forward in current debates in political science on how to assess paradigmatic changes in the global political structure, such as new forms of horizontal steering and regulation through various actors and institutions. It is appropriate to refer to perspectives on globalisation, neo-institutionalism and governance in order to answer the question in how far these concepts are considered useful in obtaining theoretically informed hypotheses concerning the role of AFC. Moreover, the method-oriented question of which specific indicators and factors help with identifying and tracing AFC's role will be answered. Finally, the analysis-oriented question of what conclusions can be drawn, with regard to both the theoretical frameworks for further sport policy research and the future development of Asian football, will be dealt with.

In consideration of the theoretical framework, three roles are suggested (see Chapter 2). The role of AFC as a 'glocal player' refers to concepts of globalisation and calls for focusing not just on the historical context in which AFC developed and established itself, its political location in an Asian and global context, and the inherent tensions and conflicts – as well as its authority – in terms of homogenisation and regulation, but also on how far AFC is affected by feedback mechanisms and other driving forces. The proposed role of AFC as 'corporate actor' draws attention to the institutional context it operates in, its function as an organisation which acts independently, and the importance of individual actors within decision-making processes and policy formulations. The governance-oriented role as a 'skilful negotiator' requires an examination of how AFC interacts with other actors in the sense of a horizontalised sport policy system.

The analysed data consists of academic literature, online sources, press documents, archive material and interviews. Field trips to AFC's and FIFA's headquarters provided the opportunity to gather primary sources and conduct semi-structured interviews with members of staff. In addition, former employees, journalists, academics and representatives of other associations or AFC-related actors functioned as interviewees. The historical period which was examined begins in 1954 when AFC was inaugurated and ends in 2011 with the ruling against Mohamed Bin Hammam at the Court of Arbitration for Sport, followed by his provisional replacement and major administrative and strategic restructuring plans within AFC. Selective updates were, however, made throughout the years 2012 and 2013 with regard to major decisions and landmarks in order to maintain a considerable scope of currency and a basis for pointing out possible future developments and areas of research.[1]

6 Introduction and contextualisation

Note

1 While taking historical developments into account, Chapters 5–7 mostly provide a synchronic analysis based on the latest data that was obtainable in this period. Office holders, statutory matters and policy outputs, for example, are therefore referred to in the present tense even if they have since changed or been amended.

References

Andreff, W., and Szymanski, S. (eds). (2006). *Handbook on the economics of sport*. Cheltenham: Edward Elgar.

Armstrong, G. J., and Giulianotti, R. (eds). (1999). *Football cultures and identities*. Basingstoke: Macmillan.

Asian Football Confederation. (2002). *AFC XXth Congress 2002: Agenda*.

BBC Sport. (2004). 'Asian football looks to future.' Retrieved from http://news.bbc.co.uk/sport2/hi/football/world_football/3704023.stm on 10 August 2010.

Beech, J., and Chadwick, S. (eds). (2004). *The business of sport management*. Harlow: Pearson Education.

Darby, P. (2002). *Africa, football and FIFA: Politics, colonialism and resistance*. London: Cass.

Darby, P., Sugden, J. and Tomlinson, A. (1998). 'FIFA versus UEFA in the struggle for the control of world football.' In A. Brown (ed.), *Fanatics! Power, identity and fandom in football* (1st edition, pp. 11–31). London: Routledge.

Dimeo, P., and Mills, J. (eds). (2001). *Sport in the global society: Soccer in south Asia: Empire, nation, diaspora*. London: Frank Cass Publishers.

Dobson, S., and Goddard, J. (2001). *The economics of football*. Cambridge: Cambridge University Press.

Duerden, J. (2011). 'Asian heads turned west.' Retrieved from http://soccernet.espn.go.com/columns/story/_/id/931581/john-duerden-reflects-on-a-growing-transfer-trend-in-asia?cc=4716 on 5 October 2011.

Ehrmann, J. and Klappenbach, M. (2011). 'Nie war der ferne Osten näher – Asiaten in der Bundesliga.' Retrieved from www.11freunde.de/bundesligen/135561 on 17 September 2011.

Eisenberg, C., Lanfranchi, P., Mason, T. and Wahl, A. (2004). *100 years of football: The FIFA centennial book*. London: Weidenfeld & Nicolson.

Fanizadeh, M., Hödl, G. and Manzenreiter, W. (eds). (2002). *Journal für Entwicklungspolitik Ergänzungsband: Vol. 11. Global players – Kultur, Ökonomie und Politik des Fußballs* (1st edition). Frankfurt am Main: Brandes & Apsel.

Gammelsæter, H. and Senaux, B. (eds). (2011). *Routledge research in sport, culture and society, Vol. 3: The organisation and governance of top football across Europe: An institutional perspective*. London: Routledge.

Garcia, B. (2008). *The European Union and the governance of football: A game of levels and agendas*. Dissertation, Loughborough University.

Giulianotti, R. (2007). *Football: A sociology of the global game* (reprinted). Cambridge: Polity Press.

Giulianotti, R. and Robertson, R. (2007). 'Recovering the social: Globalization, football and transnationalism.' *Global Networks*, 2(7), 166–186.

Giulianotti, R. and Robertson, R. (2009). *Globalization and football* (1st edition). Los Angeles: SAGE.

Groll, M. (2007). 'Wir sind Fußball: Über den Zusammenhang zwischen Fußball, nationaler Identität und Politik.' In J. Mittag and J.-U. Nieland (eds), *Das Spiel mit dem Fußball: Interessen, Projektionen und Vereinnahmungen* (pp. 177–189). Essen: Klartext Verlag.

Guttmann, A. (1994). *Games and empires: Modern sports and cultural imperialism.* New York: Columbia University Press.

Hamil, S., Michie, J. and Oughton, C. (eds). (1999). *The business of football: A game of two halves.* Edinburgh: Mainstream.

Henry, I. (2007). 'Globalisation, governance and policy.' In I. Henry (ed.), *Transnational and comparative research in sport: Globalisation, governance and sport policy* (pp. 3–21). London: Routledge.

Holt, M. (2009). 'UEFA, governance, and the control of club competition in European football: A report funded by the FIFA João Havelange Research Scholarship.' Birkbeck Sport Business Centre report, Vol. 2, No. 1, January.

Hong, F. and Mangan, J. A. (2003). 'Will the "Iron Roses" bloom forever? Women's football in China: Changes and challenges.' *Soccer & Society*, *4*(2–3), 47–66.

Hopf, W. (ed.). (1998). *Sport: Vol. 15. Fußball: Soziologie und Sozialgeschichte einer populären Sportart* (3rd edition). Münster: Lit.

Horne, J. and Manzenreiter, W. (eds). (2002). *Japan, Korea and the 2002 World Cup.* London: Routledge.

Houlihan, B. (2003). 'Politics, power, policy and sport.' In B. Houlihan (ed.), *Sport and society. A student introduction* (1st edition, pp. 28–48). Los Angeles: SAGE.

Kramer, D. (2005). 'The road to success cf Asian football'. Unpublished lecture at Sportschule Hennef, Germany.

Lee, S. (2003). 'Moving the goalposts: The governance and political economy of world football.' In A. Budd and R. Levermore (eds), *Sport and international relations* (pp. 112–128). London: Frank Cass.

Lösche, P. (2010). 'Sportpolity, Sportpolitics und Sportpolicy als theoretische Annäherung an eine Sportpolitikwissenschaft.' In W. Tokarski and K. Petry (eds), *Handbuch Sportpolitik* (pp. 12–29). Schorndorf: Hofmann.

Majumdar, B. and Bandyopadhyay, K. (2006a). *A social history of Indian football: Striving to score: Sport in the global society.* New York: Routledge.

Majumdar, B. and Bandyopadhyay, K. (2006b). *Goalless: The story of a unique footballing nation.* New Delhi: Viking.

Manzenreiter, W. and Horne, J. (eds). (2004). *Football goes east: Business, culture and the people's game in China, Japan and South Korea.* London: Routledge.

Manzenreiter, W. and Horne, J. (2007). 'Playing the post-Fordist game in/to the Far East: The footballisation of China, Japan and South Korea.' *Soccer & Society*, *8*(4), 561–577.

Mittag, J. and Nieland, J.-U. (eds). (2007). *Das Spiel mit dem Fußball: Interessen, Projektionen und Vereinnahmungen.* Essen: Klartext Verlag.

Murray, B. (1995). 'Cultural revolution? Football in the societies of Asia and the Pacific.' In S. Wagg (ed.), *Sport, politics, and culture: Giving the game away. Football, politics, and culture on five continents* (pp. 138–162). London, New York: Leicester University Press.

Redeker, R. (2008). 'Sport as an opiate of international relations: The myth and illusion of sport as a tool of foreign diplomacy.' *Sport in Society*, *11*(4), 494–500.

Samuel, P. M. (AFC General Secretary 2007–2009) interviewed by B. Weinberg, 5 August 2008, Kuala Lumpur.

Schulze, B. (2002). *Sportverbände ohne Markt und Staat: Eine empirische Untersuchung der Spitzenverbände in Uruguay*. Münster: Waxmann.

Strünck, C. (2007). 'Der Deutsche Fußball-Bund als Interessenverband.' In J. Mittag and J.-U. Nieland (eds), *Das Spiel mit dem Fußball: Interessen, Projektionen und Vereinnahmungen* (pp. 191–202). Essen: Klartext Verlag.

Sugden, J. and Tomlinson, A. (1997). 'Global power struggles in world football: FIFA and UEFA, 1954–74, and their legacy.' *International Journal of the History of Sport*, (14), 1–25.

Sugden, J. and Tomlinson, A. (1998). *FIFA and the contest for world football: Who rules the peoples' game?* Cambridge: Polity Press.

Sugden, J. and Tomlinson, A. (1999). *Great balls of fire: How big money is hijacking world football*. Edinburgh: Mainstream.

Sugden, J. P. and Tomlinson, A. (eds). (2002). *Power games: A critical sociology of sport*. London: Routledge.

Sugden, J. P. and Tomlinson, A. (2003). *Badfellas: FIFA family at war*. Edinburgh: Mainstream.

Wagg, S. (ed.). (1995). *Sport, politics, and culture: Giving the game away: Football, politics, and culture on five continents*. London, New York: Leicester University Press.

Weinberg, B. (2012). '"The future is Asia"? The role of the Asian Football Confederation in the governance and development of football in Asia.' *The International Journal of the History of Sport*, 29(4), 535–552.

2 Analysing the Asian Football Confederation
Theoretical perspectives and approaches

Sport has often been and is still regarded as politically neutral by most actors involved in its organisation and implementation processes. Paradoxically, sport is often portrayed as genuinely apolitical, stressing its distinct sportive character (in order to prevent sport from being negatively connoted), while other actors claim it to be facilitative in terms of integration, inclusion and nation-building (in order to promote its supposedly exclusively positive attributes). In comparison to these somewhat simplifying myths, it has to be stated that, first, a relationship between sport and politics does in fact exist and that, second, this relationship is multi-layered and multi-dimensional (Woyke and Delschen, 2006). The following section shall therefore depict how sport is intertwined with political aspects – both directly ('kleine Politik') and indirectly ('große Politik') – and how this phenomenon has been explored from a (sport-)political science perspective (Güldenpfennig, 2011).

The theoretical and methodical approaches to examining sport politics and policies remain highly differentiated depending on one's definition of 'politics'. Basically, two positions have emerged. The first sees politics as exclusively associated with governmental actions and is therefore primarily concerned with decision-making processes that state actors are involved in, as well as specific policy analyses. These studies mainly deal with public sport policies and political systems, which are determined through governmental actors. Therefore common topics of interest are the regulation or prohibition of certain types of sport, sport as training for war, sport as a vehicle for social integration, sport and national identity, the support of elite sport as a means of enhancing national self-image, sport as an economic motor or sport as a diplomatic tool (Houlihan, 1994, 1997, 2003, 2006).

The second position draws upon the recognition that the organisation and management of the sport sector in an era of deregulation and liberalisation is increasingly characterised through the inclusion of non-state actors, such as international NGOs and global enterprises (Hödl, 2002; Sage, 2006). The notion of citizens governing themselves due to the incremental dissolution of the so-called strong state is gaining in significance (Spitaler, 2009: 274). In the context of a wider definition of politics and power, attention has also been drawn to analytical categories such as class, race and gender, and the way power is being produced and generated through political discourses and ideologies (Hargreaves, 1994; Hong and Mangan, 2003; Fozooni, 2004; Majumdar and Bandyopadhyay, 2005; Hayhurst, 2009).

Making the important distinction between sport as as '*Tätigkeitsystem*' (a system of activity) and sport as '*institutionelles System*' (an institutional system), Güldenpfennig (1981: 24–25) refers to the prior as exclusively relating to the actual physical activity according to specific rules and without any political reference, and to the latter as the sum of all social providers of sport, its organisations, and interrelations with other social and political institutions. It is this area where sport becomes political due to the fact that, on the one hand, these organisations and their issues of power and spheres of influence constitute significant parameters, and on the other hand because the actors of the sport sector express and exercise their interests towards or together with the state (Lösche, 2010: 13).

In a similar vein, Schmid (1998: 3) has classified the mediation of interests through associations into three dimensions: first, the polity dimension, which refers to the institutionalisation of mediation through norms, values and rules of procedure; second, the politics dimension, which deals with the process of political decision-making and creating power and spheres of influence in the context of enforcing interests; and third, the policy dimension, which refers to the outcomes of mediation processes as binding for society.

German sport sociologists have frequently put an emphasis on the autonomy and self-administration/regulation of the sport sector. According to this view, sport is interpreted as a non-political functional system which is defined through its self-regulating capacity. In this sense, political interventions lead to dysfunctional effects or an instrumentalisation of sport (Rütten, 1996). This view has its limitations in assessing the symbiotic relationship (interdependence) between professional sport and the media, as well as aspects of sport-specific regulation through governmental actors. In addition, it seems problematic to neglect the fact that the sport sector is significantly dependent upon the mobilisation of resources through the political system. Thus more importance needs to be given to interest groups' and associations' role as bargaining institutions between the state and actors such as clubs and players. Also, the fact that sport associations' and federations' political activities are constituted through the exercise of power by officials and delegates substantiates considering sport organisations as corporate actors that are interested in maintaining their powers, autonomy and provision of resources. This indicates the necessity for employing political science perspectives in sport studies, while being cautious of not reducing sport to its political dimension (Meier, 2005: 10–13).

Viewing political science as an integrative science that borrows theoretical frameworks and methical tools from other disciplines and adapts them, scholars of international relations and foreign policy analysis have used the concept of 'role' in order to assess the behaviour of states and international organisations. Harnisch *et al.* (2011: 1–2) argue 'that it is meaningful as an analytical tool to explore both national foreign policy behaviour and developments in international relations as a social system.' They go on to say that roles, as 'the notions of actors about who they are, who they would like to be with regard to others, and how they therefore should interact in (international) social relationships', therefore serve as a useful tool at the intersection between the analytical levels of actors and the system. The authors draw attention to the fact that roles undergo constant recreation, redefinition and modification. Simultaneously they point to the 'high degree of path dependency' that some roles may display. Harnisch (2011: 8) acknowledges the 'systemic dynamics of role change' and the importance of analysing 'complex role sets', arguing that 'roles in international relations cannot be thought of or theorized about without reference to other roles and a basic recognition through society'.

In accordance with an understanding of sport organisations as political entities, Sugden and Tomlinson (1998: 299ff.) have offered a theorisation of how to view the role of FIFA. They presume that it is useful to consider the organisation as an example of the increase of international non-governmental organisations that are characterised by having 'global remits' but are not 'accountable to any particular national government or governments', and are therefore distinct from international governmental organisations such as the United Nations or the European Union. This observation resonates with how Houlihan (1994: 55ff.) and Eisenberg (2006: 61) have defined the IOC and FIFA respectively as INGOs or transnational actors that, as pointed out by Houlihan, 'operate across national boundaries and seek access to a number of sovereign states'. They also agree on how this status is associated with issues of globalisation, politics and business. Eisenberg (2006: 55) refers to FIFA as having the 'role of a global player in the relationship between sport and politics' and 'as a business against the background of its football development programmes, which have made the world soccer federation a most effective International Non-Government Organisation.'[1] In this sense, Sugden and Tomlinson (1998: 312–313) reason that 'FIFA can be viewed both [as] a transnational body that promotes globalization (and transnational capitalism) and as a locus for resistance'. Adding that 'FIFA and the confederational substructure' can be exploited 'as sites for the construction of personal fiefdoms', they conclude that 'FIFA's expansion could not have been achieved without the considerable support of major commercial sponsors'.

The roles of international federations or confederations such as AFC are therefore inevitably linked with aspects of globalisation, transnationalism, postcolonialism, governance, neo-institutionalism and commercialisation. By analogy with various political science perspectives on how to assess INGOs in the context of international relations, integration and cooperation in general, it is obvious that the theorisation of international sport organisations features a similar degree of diversity.

Globalisation

The concept of globalisation has been dealt with in various disciplines and therefore the academic discourse on the topic is manifold. Since the 1990s the term has been increasingly discussed by sociological, economic and political science scholars, while some claimed that it does not necessarily describe new historical developments or processes. Indeed, as outlined above, issues such as the internationalisation of politics and the growing interlacing of various non-governmental and governmental actors beyond traditional boundaries of the nation state have been addressed, for instance through interdependence and regime theories. Also, earlier sociological theories have dealt with modernisation and civilisation processes including aspects such as the expansion of so-called western culture (Elias, 1994).

The end of the Cold War constituted an impetus for assessing a new global constellation strongly associated with the capitalist idea of a free global market. Hence, globalisation came to be mainly understood in economic terms while not being reduced to it. In fact, cultural norms and values, and social and political manifestations, are also relevant aspects of globalisation (Meyer, 1997). Overlapping all disciplines is the recognition that globalisation has to do with relations between local, regional, national, transnational and international phenomena, while the effects, outcomes, manifestations and reciprocities of these processes remain vigorously debated.

Broadly speaking, three major streams of globalisation have emerged. The hyperglobalisers view globalisation as a unidirectional process resulting in the creation of a global society in which the powers of the nation state are diminished, thereby dissolving national identities and giving way to cosmopolitan identities. The second stream are sceptical of this theory and consider the presumption of the fall of the nation state to be a myth, and also express scepticism towards the supposed homogenising effects of globalisation on national identities and nationalism (e.g. Huntington, 1997). The transformationalists (e.g. Giddens, 1990; Held, 1999; Robertson, 1990) consider contemporary globalisation a unique order but identify it as a not entirely new development, while emphasising its discontinuity and openness. As for the role of the nation state, they contend that it is not necessarily weakened but rather challenged by global forces, requiring adaptation and reconstruction processes (Hargreaves, 2002: 25–27; Lee, 2003: 113).

Several scholars have made attempts to develop differentiated definitions of globalisation by drawing attention to the frequent misuse of the term. Robertson (1990: 15ff.). for example, states that:

> While there is rapidly growing interest in that topic, much of it is expressed very diffusedly and there is considerable danger that 'globalization' will become an intellectual 'play zone' – a site for the expression of residual socio-theoretical interests, interpretive indulgence, or the display of world-ideological differences. [...] Rather, I am insisting that both the economics and the culture of the global scene should be analytically connected to the general structual and actional features of the global system. [...] Globalization as a topic is, in other words, a conceptual entry to the problem of world order in the most general sense – but, nevertheless, an entry which has no cognitive purchase without considerable discussion of historical and comparative matters.

He advocates a 'temporal–historical path to the present circumstance of a very high degree of global density and complexity' (Robertson, 1990: 26) by establishing a five-phase (the germinal, incipient, take-off, struggle-for-hegemony and uncertainty phases) 'skeletal sketch' on globalisation, which has later been adapted to sport, and football in particular, by Giulianotti and Robertson (2007, 2009).

Touching upon the assumption that traditional national boundaries and both international systemic structures and activities have to be thought of differently, the notion of transnationalism or transnational relations is often utilised for referring to conditions under which non-governmental actors emerge as independent actors, as well as ones influencing the actions of states (Grande and Risse, 2000). According to Nye and Keohane (1971: 332, 335), transnational interactions can be described as 'the movement of tangible or intangible items across state boundaries when at least one actor is not an agent of a government or an intergovernmental organization'. In this sense transnational relations 'include the activities of transnational organizations [...], even when some of their activities may not directly involve movements across state boundaries and may not, therefore, be transnational interactions as defined above'. In sociological terms, meanwhile, transnationalism is understood 'as referring to processes that interconnect individuals and social groups across specific geo-political borders' (Giulianotti and Robertson, 2007: 170).

The concept of globalisation has been frequently dealt with in relation to sport. In particular, sport sociologists have looked at the interdependencies between sport development and the increase of transnational corporations, the deregulation of labour markets, the emergence of a postmodern consumer culture, and an unprecedented growth in global mobility, images and ideologies (Whannel, 2008: 198). Consequently, topics such as the spread of western and especially British sports culture (Elias and Dunning, 2008; Eisenberg, 1997; Mangan, 2001; Guttmann, 1994), the establishment and significance of international competitions and federations (Eisenberg, 2007), the relationship between sports and media (Horne and Manzenreiter, 2002) and the formation of a global sports industry, including aspects of merchandising and marketing (Slack, 2004; Manzenreiter and Horne, 2007) and athlete migration (Bale and Maguire, 1994), have been studied.

What can be identified are profound changes deriving from globalisation, its effects on nations, and hence the necessity of looking beyond a Westphalian state system. This does not, however, mean that these developments can simply be equated with homogenising ramifications (Maguire, 1999). Instead, it is more viable to work with a conceptualisation of globalisation as a multi-dimensional and uncertain process consisting of patterns of 'transnational or global economic and technological exchanges, communication networks' and migration that lead 'to a form of time–space compression' and 'a greater degree of interdependence' (Maguire, 2006: 356). Reflecting on the perils of maintaining binary oppositions in globalisation studies – i.e. universalism vs. particularism, homogenisation vs. differentiation, integration vs. fragmentation, centralisation vs. decentralisation, juxtaposition vs. syncretisation – Maguire (1994: 402) argues that:

There is no *single* global flow; in the interweaving of global scapes, disjunctures develop and cause a series of diverse, fluid, and unpredictable global conditions and flows. Competing and distinctive cultures are thus involved in an infinitely varied, mutual contest of *sameness* and *difference* across different figurational fields. Rejecting the idea of some global cultural homogeneity does not, however, mean accepting the idea of some haphazard, unstructured growth in global cultural diversity. In highlighting issues of homogeneity, and the mutual contest of sameness and difference in global cultural flows, the analysis can be developed with reference to the twin figurational concepts of diminishing contrasts and increasing varieties.

Building upon Elias' (1994) figurational approach and Robertson's (1998) considerations, and through revising and extending these, Maguire believes: 'Globalization is best understood as a balance and blend between diminishing contrasts and increasing varieties, a commingling of cultures and attempts by more established groups to control and regulate access to global flows.'[2] In this context, Maguire has developed the idea of a five-phase sportisation process in which there has occurred an interweaving of the global and the local, 'an amalgamation of the "Western" and the indigenous patterns'. Accordingly 'the fourth phase [author's note: 1920–1960] of sportization clearly involved an elaborate political economy in which hegemonic control of sport lay with the "West", [although] control was never complete'. In fact, the latter stages of the fourth and then the fifth phase 'were also characterized by the rise of non-Western nations to sporting prominence [...] however, [...] still [...] on western terms, for they do so through "Western" sports.' Detecting a trend towards a 'creolization of sports cultures', he stresses the need for more 'detailed empirical case studies' in particular in relation to 'the "reach" and "response" of global flows on local cultures' (Maguire, 2006: 358–366).

Referring to these aspects, their analytical implications and the difficulty of measuring diversity, Houlihan (2009; 2010) has taken a look at the mechanisms of international influence on domestic elite sport policy. He argues that 'the mechanisms vary in relation to the locus of initiative, the basis of engagement, the key relationships and the nature of power relationships', as well as that in some countries 'there is a dual process in operation of domestically initiated policy learning and non-domestically initiated policy harmonisation through policy regimes'. The intended harmonisation through international regimes suggests that 'the significance of the state may be balanced or replaced by that of non-governmental organisations such as the International Olympic Committee (IOC) and the international federations'. However, Houlihan concludes that the impact of such regimes 'will be uneven; it is generally the weaker states whose domestic policy is more likely to be influenced.' (2009: 51–66) He also expresses his doubts as to 'whether international sport organisations have become more politically astute' considering 'that the relative autonomy of sport organisations has declined' due to the fact the supranational actors such as the European Commission, the European Parliament and the European Court of

Justice have had increasing interest in and found ways of regulating sport, for example through the Bosman ruling (2010: 59–60).

Bearing in mind that globalisation revolves around the idea of a world that is increasingly interconnected, football 'may be understood as a metric, a mirror, a motor, and a metaphor of globalization'. This does not imply that globalisation has led to overall dominance of western culture. In fact, the complexity of global and local relations could be better described through the term 'glocalisation', which refers 'to "real world" endeavours to recontextualize global phenomena or macroscopic processes with respect to local cultures' (Giulianotti and Robertson, 2009: xii, 2). Despite the fact that 'football's homogenization is evidenced transnationally by isomorphic forms and institutional structures', heterogeneous manifestations can be found in 'various national sporting codes, such as "American football" or [...] Gaelic football in Ireland'. Another example is that 'at club level, the political hegemony of the richest Western European clubs has resulted in the relative disconnection of sides based in smaller nations' while at the international level 'developing football continents of Africa or Asia have been in recurring conflict [with] FIFA since the mid-1960s over their restricted allocation of qualifying places for the World Cup finals' (Giulianotti and Robertson, 2007: 168–172).

In parallel to Maguire's sportisation approach, and by analogy with Robertson's phase model, Giulianotti and Robertson identify five phases of a global history of football. The first and second are the 'germinal' and 'incipient' (up to the 1870s) phases, which are relevant in terms of locating the origins of both ancient ball-kicking games, such as the Chinese *cuju* and the Japanese *kemari*, and modern association football as codified by the British Football Association in 1863. The third phase, titled 'take-off', ranges from the 1870s to the mid-1920s and is characterised by the internationalisation of the modern game (e.g. in British colonies and with the founding of FIFA in 1904). The fourth phase, named 'struggle-for-hegemony', then continued from the mid-1920s to the late 1960s and is mainly associated with the hosting of the first World Cup tournament in 1930, the emergence of professionalism and the expansion of football's global system, resulting in the establishment of continental governing bodies such as UEFA and AFC (both established in 1954) and the instrumentalisation of sports for political and ideological purposes. The final phase, called 'uncertainty' and which spans the late 1960s to the early 2000s, is characterised by two major international developments, namely the expansion of the political influence of non-European football societies, finding its expression in the successful FIFA presidential candidacy of João Havelange in 1974, and second the mass mediatisation of international football. Furthermore, the authors argue that 'since 1990, football's political management became both more complex and more closely integrated with the broader international system' which has led to a crystallisation of:

> a wider array of institutional actors, such as governing bodies (global, continental, and national), media and merchandise TNCs, organized labour

(especially the professionals' world union FIFPro), player agents, international federations (especially the EU), the world's top clubs, and supporters' organizations.

(Giulianotti and Robertson, 2009: 5–25)[3]

In line with Bourdieu's ideas, Horne and Manzenreiter (2004: 4–15) point out that, for assessing relations between sport and globalisation, it is certainly important to consider all of these issue but that, in contrast to Robertson's multi-dimensional conception, 'the relative significance and relation of each of the separate, but not separable dimensions of globalization, revolves around the economic significance for the other dimensions'. However, this does not mean that they regard globalisation as a unidirectional transformation process, but rather as a hybrid of homogenisation and heterogenisation mainly driven by economic policies and business strategies. Specifically, they identify an 'ongoing increase in the value of genuinely global sport properties, including athletes and players themselves, and the convergence of economic power in sport ownership' which can be witnessed for instance in the exponential increase of FIFA's financial activities and the trend towards exploiting 'the economically vibrant East Asian region' as 'a preferential destination for marketing managers and promotion tours'. Stressing 'the need for comparative studies of sport in different cultural (political, economic, commercial and policy) contexts', they conclude that 'hardly any Western product has proved to be more successful, pervasive, and persistent than sport', although it took 'more than 70 years before Asia was the first time conceded the right to host the Football World Cup Finals.'[4]

The role of AFC as 'glocal player'

Based on the theoretical underpinnings, the academic discourse on globalisation – which understands globalisation as a multi-directional time–space compression encompassing political, economic, social, cultural and historical dimensions, and as revolving around tensions between the global and the local as well as homogenisation and heterogenisation – can be presumed to have a considerable impact on the assessment of the role of AFC in Asian and global football politics.

In relation to this work, it should be borne in mind that the fundamental considerations made by Maguire (2006) and Houlihan (2009) on issues of harmonisation and heterogenisation, as well as Giulianotti and Robertson's (2007) notion of glocalisation, constitute pertinent anchor points for an analysis of AFC's relations with, and the impact of its policies (harmonisation) on, specific member associations. The Vision Asia football development programme, for instance, could be viewed as an international sport development regime. These ideas also refer to the relationship between modern football as a western sport and how it is responded to and perceived in an Asian context (e.g. the issue of the dominance of European football leagues in the media and the increasing migration of players to Europe). They tie in with questions of how far AFC may understand and image itself as a developer of a distinct Asian football culture and system (e.g. the significance and promotion of Futsal), to what degree EU sport policy may have indirect effects on regulations in Asia, and whether respective regional networks such as the Association of Southeast Asian Nations may gain interest in sport policy issues. Furthermore, the phase models developed by Maguire (2006) and Giulianotti and Robertson (2009) seem particularly viable in relation to assessing the historical development of football in Asia.

Accordingly, it is suggested that the AFC could be defined as a 'glocal player' with a multipolar directionality, placing itself on a spectrum between promoting either a global or a local/regional (Asian) business, media and management-oriented outlook within its power relations and network constellations. The focus and intensity varies depending on what specific issues concern extra- or intra-Asian relations, as well as on the political, economic, social, cultural and historical context. Rather than reducing the perspective of this examination to one specific model of globalisation, a broad consideration of various approaches and a soft application are more appropriate with regard to analysing the proposed role. This helps to understand the vital conjunctions between globalisation, neo-institutionalism and governance, and therefore their theoretical potential for identifying the role of AFC in football politics.

The globalisation discourse is closely associated with a discussion on the emergence of governance structures, including new forms of political management and regulation through international regimes, organisations and complex networks consisting of private and public actors, as well as with an increase of academic works that reconsider institutional factors within the context of policy analyses (Grande and Risse, 2000: 236ff.; Lee, 2003: 112). 'Three types of issues arise out of the globalisation of sport', namely 'regime development', 'the relationship between emerging global institutions and existing institutions' and 'cultural political struggles arising out of the tendencies to create a global culture' (Allison, 2005: 3–4). Finally, there is a need to apply classic issues of political science analysis – 'representation, responsibility, accountability, agency-structure and, above all, power and its abuse' – to the study of the roles played by international federations and confederations such as FIFA and AFC (Sugden and Tomlinson, 2005: 26ff.). This relates to the perception that 'globalization is bound together by what Castells refers to as "the network society"' in which 'big business networks are the overriding influence in the global development and political control of world football.' (Sugden, 2002: 63–64)

Neo-institutionalism

While the theorisation of international relations and integration experienced a phase in which some studies primarily looked at behaviourist and functional aspects of driving forces, the so-called institutionalist turn in political science has initiated a reconsideration of the significance of institutions in relation to explaining patterns of agency and structure, decision-making and participation, and to policy programmes and outputs.[5] Against the backdrop of the emergence of regime theories and debate about the relation between institutions and organisations, neo-institutionalist approaches have drawn attention to the assessment of individual and complex action units within certain institutional settings and environments (Schneider and Werle, 1989: 410–415). In this sense, for example, Coleman's (1971) concern with the constellations between the interests of corporate and individual actors, and their institutional structures, rules, competences and external influences, has advanced perspectives in terms of recognising international organisations' discretion to act independently through institutional and contractual frameworks, e.g. with regard to the role of the EU as an international corporate actor (Kenis and Schneider, 1987; Schneider and Werle, 1989). Accordingly, policy analysis studies have increasingly acknowledged the importance of utilising polity as one possible dimension for explaining the procedure of politics and the formation of policies (Blum and Schubert, 2011: 34–35).

The definitions and applications of 'actor', 'institution' and 'structure' have varied depending on the locus of analysis. While there has been mutual understanding on the necessity of answering the question 'do institutions matter?' (Weaver and Rockman, 1993), and hence of locating the 'place of institutions in politics' (March and Olsen, 1984: 735), neo-institutionalist works in political science can be broadly divided into three schools of thought. Each school borrows from approaches of other disciplines but differ with respect to 'two issues fundamental to any institutional analysis, namely, how to construe the relationship between institutions and behaviour and how to explain the process whereby institutions originate or change' (Hall and Taylor, 1996: 937). These schools are rational choice institutionalism, sociological institutionalism and historical institutionalism.[6]

Focusing on the strategic behaviour of actors, rational choice institutionalism constitutes an institution-oriented version of the rent-seeking paradigm, which understands political decision-making processes as a result of interactions between actors who are driven by self-interest. This approach follows the model of the *homo oeconomicus*, according to which on the micro-level individuals act exclusively dependent upon their individual potential benefit, which is being sought through rational calculation (Meier, 2005: 18). Rational choice institutionalists assume that the relevant actors have a fixed set of preferences, 'behave entirely instrumentally so as to maximize the attainment of these preferences, and do so in a highly strategic manner that presumes extensive calculation' (Hall and Taylor, 1996: 944–945). Furthermore, they regard politics to be a series of collective action dilemmas in which actors try to attain their preferences and probably create a suboptimal collective outcome. They also assume that the behaviour of actors is not driven by impersonal historical factors, but by strategic calculus and by their expectation of other actors' behaviour. In this sense, institutions are primarily seen as dependent variables that structure these interactions through setting the agenda, providing information and generating mechanisms that in turn serve as norms and rules for strategic behaviour. In relation to the question of how institutions originate, rational choice institutionalists stress the functional value of an institution for a specific actor. Depending on the potential benefit of cooperation, the creation of an institution is therefore understood as deriving from voluntary agreements (Schneider and Janning, 2006: 88ff.; Hall and Taylor, 1996: 945).

In contrast, sociological institutionalism is based on an integrative relation between actors and institutions. Actions of individuals are viewed as determined by institutions, which are thus understood as independent variables. The binding power of institutions does not derive from individual preferences but from compliance with norms and rules. Consequently, any action and decision carried out by an actor is embedded in an institutional framework. Strategies and preferences therefore do not result from a conscious decision-making process aimed at maximising one's benefit, but they rather constitute an endogenous product of an institutional environment shaped by traditions and conventions. Thus institutions are seen as establishing symbolic and cognitive elements upon which actors explore their preferences and define their identities (Morisse-Schilbach, 2006: 276; Powell and DiMaggio, 1991). This does not imply that rational motives are regarded as entirely irrelevant; in fact, sociological institutionalists understand rational actions as socially constructed and thus conceptualise the objectives that an individual pursues in much broader terms. Concerning the aspect of how institutions originate or change, they contend that institutional practices are not advanced for efficiency-oriented purposes but due to reasons relating to social legitimacy of an organisation or its members (Hall and Taylor, 1996: 949).

Historical institutionalism compensates the other approaches' neglect of the temporal dimension, namely the historicity of institutions and political decision-making processes. As has become evident in other comparative studies of political science, institutional arrangements and interactions are characterised by a certain degree of continuity, including relevant cultural and ideological path dependencies. Whereas rational choice and sociological institutionalism conceptualise the relation between actors and institutions as either determined by actors or institutions respectively, historical institutionalism locates itself in between by attributing the time factor an important role as a variable. Historical institutionalism proceeds on the assumption that present institutions are products of singular historical constellations and critical junctures, while including elements of continuity and dynamics. In this sense, cultural path dependencies, policy legacies and unintentional consequences are viewed as important with regard to the process of institutionalisation (Meier, 2005: 20). Defining institutions basically as formal or informal norms, rules and conventions, historical institutionalists associate institutions with an organisational capacity. They tend to view the relationship between actors and institutions in relatively broad terms by eclectically taking both the cultural and the calculus approach into consideration.[7] The roles of power and asymmetrical relations of power constitute further important aspects of historical institutionalist analysis; it is assumed that institutions distribute power unevenly across groups and actors. Finally, historical institutionalists contend that institutions are not the only causal factor in explaining politics. Rather, the latter is also linked to socio-economic and ideological developments (Hall and Taylor, 1996: 942).

As has been recognised by Meier (2005: 21), all three schools are capable of delivering important insights with regard to the examination of political processes, but nonetheless they employ a highly selective focus. It is therefore very useful that, through the development of so-called actor-oriented institutionalism, a neo-institutionalist approach has been established which understands itself as integrative research heuristics reflecting theoretical preconditions and explanatory power, while also taking the assumptions of the other approaches into consideration.

This approach was created and evolved by Renate Mayntz and Fritz W. Scharpf during the 1990s and is based on a narrow definition of institutions, which are viewed as both dependent and independent variables without having a determining impact on interactions. Institutional factors are rather thought of as shaping a stimulating, enabling or restricting action context. This context is considered as consisting mainly of interactions between corporate actors, which are defined as organisations capable of acting in a societal sector that is systematically characterised by a high degree of self-organisation. While putting the focus on this type of actor, individual actions at the micro-level are also regarded as important due to the fact that the strategies of corporate actors can at times depend on non-institutionally determined actions carried out by individuals in their respective roles as members or representatives. As assumed by the approach, this implies a multi-level perspective in which the institutional framework shapes the actions of organisations, while these in turn constitute the institutional framework for the actions of its members (Blum and Schubert, 2011: 44ff.).

In contrast to institutionalist approaches that concentrate solely on political institutional arrangements, actor-oriented institutionalism aims at providing a higher degree of abstraction by contemplating all relevant institutions and actors, namely governmental and non-governmental actors and their corresponding structural frameworks. Unlike sociological institutionalism, institutions in this sense are reduced to aspects of regulation and rules. Mayntz and Scharpf (1995a: 43–48) substantiate this analytical perspective through, first, rejecting the idea of institutions as being the result of evolutionary and irreversible developments, instead advocating the notion that they can be intentionally designed and modified (dependent and independent variables), and second, through drawing attention to the presumption that institutional contexts shape and restrict actions but do not determine them. A narrow definition of institutions – which concedes that actions have status as an independent variable and is therefore able to analyse situations in which significant changes in actions occur, despite an unchanged institutional framework – is viewed as more appropriate in terms of explaining steering and regulation processes at the macro-level. Actor-oriented institutionalism delivers a two-fold perspective on actors and institutions in order to overcome the analytical dichotomy between 'structure' and 'agency': that is to say, observable actions function as proximate cause and institutional frameworks as remote cause, while numerous other factors intervene between these poles, such as non-institutional factors that influence one's orientations, relations and situations.

Mayntz and Scharpf (1995a: 49) contend that social entities such as organisations can be analysed institutionally, specifically with regard to their incorporated rules, and moreover in relation to their capability of acting as corporate actors. Following Coleman's (1974) definition, corporate actors are here understood as formally organised bodies of persons capable of acting due to the fact that they possess centralised resources, and that these resources are utilised upon either hierarchic (e.g. in companies or authorities) or majoritarian (e.g. in parties or associations) decisions.

Coleman's approach conceptualises corporate actors as creating a new reality going beyond the actual mutual agreement, thereby constituting more than a mere sum of particular interests and objectives of individual actors. Accordingly, this capability is considered a variable; however, all corporate actors consist of individual actors and it is therefore important when examining the actions of corporate actors to look at the action orientations of individuals who act within and for an organisation.

While institutional regulations play an important role, the actual scope of action derives to a significant extent from the orientations that actors have. These in turn are viewed as institutionally shaped, as well as through the constellations in which actors find themselves. In addition, and as opposed to rational choice institutionalism, strategic objectives and aims are not regarded as purely driven by rational calculus but rather as conditioned by context-free attributes of the actor that derive from socialisation and general experience. Mayntz and Scharpf (1995: 53–57) therefore suggest a classification of actions orientations, which considers egocentric as well as system-related orientations and cognitive as well as motivational aspects of actions. In addition, they refer to interactive orientations that individual actors develop in the course of interpreting relations between different actors.

Due to the fact that actors always act in concrete situations, the latter form another important point in analysing systems of action. The specific requirements that a situation bears are viewed as activating latent action orientations, and consequently a concern and a participation of actors. In this sense, situations generate a set of possible options which are only partially influenced by institutional regulations, while also being deriving from non-institutional factors such as the actually available resources of actors and networks. All immediate action-relevant aspects of a situation are considered environmental aspects that are filtered and selected through own perceptions and accomplished by utilising the available resources. Beyond these subjective perceptions (constructivist understanding of constitutive interaction situations), a 'real situation' is assumed that can be identified by a hypothetical omniscient observer who could decide upon the success or failure of an action (Schneider and Janning, 2006: 95).

The resolution of problems is almost never an exclusive issue of one single actor, but in fact typically of interactions in a certain constellation between various actors with interdependent options for acting. Looking at two types of actors – individual and corporate – can thus be of limited value, taking into consideration that between these actors, as well as between members of different organisations, overlapping resource transfers and group formation processes can occur, which again can lead to the establishment of new relations, networks and changes in individual or collective autonomy. Mayntz and Scharpf (1995: 60–65) therefore refer to the term 'governance', which is thought of as characterising complex networks and collective organisations of interests, thereby constituting an own factor in describing processes of steering and self-organisation in and between organisations. These forms of governance, namely complex modes of coordination such as anticipative or reciprocal adjustments or negotiation, complement traditional modes of steering such as hierarchy, solidarity and the market. The increase of complex relations of interactions may presumably lead to bypassing conventional rules and regulations, depending on whether the necessary resources can be mobilised or not.

With regard to studying sport policies and politics, Meier (2005) has employed an institutionalist perspective in his work on the political regulation of football in Germany, while Gammelsæter and Senaux (2011: 3–6) have utilised this theoretical framework for their edition on football governance structures in Europe. Drawing upon institutional organisation theory and focusing on institutional logics, national embeddedness, and organisational fields and transnational change, the latter presume that football can be analysed '(1) as an institution in itself [...] (2) as a mosaic of national systems [...] or (3) as a community of organisations whose interweaving basically rests on cognitive-cultural foundations that span sectors and national borders'. In this sense, they view actors (individuals, groups or organisations) as exposed to and restricted by certain institutional and social guidelines or contexts, while nevertheless being capable of adapting to or even changing their values and norms, as well the institutions themselves (e.g. the Bosman case). Accordingly, decisions and outcomes are seen as resulting from the interplay between institutional structures and agency. Actors in football are simultaneously attached to and depend on contradicting institutions such as the civil sector, the market, the state and the sport itself. Sport organisations are therefore regarded as being influenced by contending logics of different institutions and structures, which in turn are spatially determined.

The role of AFC as 'corporate actor'

Based on the theoretical underpinnings, it can be presumed that neo-institutionalist approaches, and in particular actor-oriented institutionalism, have a considerable impact on the assessment of the role of AFC in Asian and global football politics. These approaches emphasise that 'institutions do matter' with regard to understanding the relation between interactions carried out by individual or corporate action units in specific constellations, as well as situations and political decision-making processes resulting in specific policy outputs and their respective repercussions.

In relation to this study, neo-institutionalist approaches contain a high degree of analytical potential because they have produced a reconsideration of the significance of institutions in relation to explaining patterns of agency and structure, decision-making and specific policies. Hence they provide a basis for understanding the institutional dimension of AFC, while acknowledging the relation between this dimension and different types of actors and interactions within an institutional environment. Historical institutionalism adds a temporal aspect to these dimensions by attributing the time factor an important role as a variable with regard to identifying patterns of continuity and change. It is consequently necessary to look at the historical developments of and within AFC in the broader context of the history of football in Asia.

Of particular interest to this study is Mayntz and Scharpf's actor-oriented institutionalism. Its two-fold perspective on actors and institutions provides theoretically founded integrative research heuristics that help explain and reconstruct interactions that take place within an institutional framework between actors in certain situations and constellations (Scharpf, 2000). The definition of corporate actors as formal organisations capable of acting while consisting of individual actors is particularly viable with respect to understanding actions of and within AFC, as well relations with other actors such as national member associations and governments. The narrow understanding of institutions, which are considered both dependent and independent variables, allows analysing AFC institutionally, namely with regard to its incorporated rules, while simultaneously relating to its capability of acting as a corporate actor. Accordingly, it could be assumed that certain institutional frameworks and logics (e.g. the economy, politics, culture or sport itself) shape, enable and constrain the actions of AFC, while the latter in turn is part of the institutional framework guiding the actions of its members. Moreover, it creates ground for tracing institutional changes and modifications initiated by specific actors. Finally, actor-oriented institutionalism and its notion of constellations leading to informal networks affiliates with governance concepts.

Accordingly, AFC could be defined as a 'corporate actor' that is capable of acting as an organisation within a dynamic institutional framework, while simultaneously consisting of individual actors and engaging in relations with other actors such as other federations or governments, thereby shaping complex constellations all of which have to be understood in a historical perspective. In this sense it can be assumed that the institutional framework shapes the actions of AFC, while the latter belongs to the institutional framework that influences the actions and decisions of its individual representatives and members, who in turn can initiate institutional changes or modifications.

In methodological terms, Meier (2005: 23–26) has pointed out the issues actor-oriented institutionalism faces when it comes to empirical analyses. He refers to the 'small-N' problem which impedes multivariate procedures of testing hypotheses, while mentioning that actor-oriented institutionalism does not primarily intend verifying or falsifying hypotheses in the first place. Rather, actor-oriented institutionalism tries to generate causal reconstructions and explanations of specific cases that may reach out in terms of identifying a manifested consistency or 'historical trajectory'. Finally, Meier refers to a modular procedure of reconstruction that can help overcome the dilemma between the complexity of the case and the challenge of generalisation by drawing upon causal and social mechanisms (neither statistical nor experimental) that trace processes and thereby ultimately result in an inductive reconstruction that can merely be seen as an interim answer.

Against the backdrop of synthesising three theoretical approaches in a soft way, it is not intended here to either conduct a detailed quantitative network analysis or game theoretical examination of interactions, but it is rather important to have cognisance of the full complexity of a simultaneous institutionalist and actor-oriented research approach. It is not necessary to investigate actor-specific orientations when institutionally ascribed tasks are sufficient enough to explain certain processes and vice versa (Mayntz and Scharpf, 1995a: 67). In fact, this study aims at resorting to several aspects of each theoretical approach in order to at least partially explain the complexity in which AFC performs its role.

Governance

The term 'governance' has increasingly found its way into academic debates within the last 20 years, producing various definitions and understandings depending on a discipline's perspective as well as the particular subject examined. In fact, the huge increase of interest in utilising this concept has been met with both scepticism and appreciation. Hence it has been viewed on a scale between regarding it as either a mere fashionable term or a useful scientific concept.

As pointed out by Benz (2004: 12–13), it is virtually impossible to produce one unitary definition of governance due to the diversity of issues that can be covered by the term. A definition therefore depends on the scientific perspective. He explains the broad appeal of governance by referring to the reciprocal relation between producing new terms in order to describe changes in reality and how reality responds to the emergence of these new descriptions, while stating that governance does not necessarily describe entirely new phenomena. In lexical terms, governance is defined as 'the action or manner of governing a state, organisation, etc.' (Oxford Dictionary Online), thereby putting emphasis on how something is governed. Hence it implies procedural, structural, functional and instrumental aspects of governing, steering and coordinating. From a political science perspective, governance could thus be viewed as the big picture consisting of polity, politics and policy (Benz, 2004: 15).

Governance was not originally shaped by political science. It rather first emerged in economics (Williamson, 1979). Referring to mechanisms of coordination and institutional regulations within a company, the term 'corporate governance' has gained significant popularity in order to describe organisational structures or relations between shareholders and managers. Furthermore, ever since the 1980s the concept of so-called 'new public management' has been dealt with in this context. Relating to the normative dimension of so-called good governance (which has been initially defined by the World Bank to describe efficient and transparent administrative and governmental processes), it is either used to refer to 'introducing private sector management methods to the public sector' (managerialism) or 'to introducing incentive structures (such as market competition) into public service provision' (new institutional economics) (Rhodes, 1996: 654–655). Generally speaking, governance in economic terms is therefore used to conceive the meaning of rules and regulations, structures, processes, methods of implementation and coordination, and aspects of power, thereby constituting basic links with how the term is used in political science.

In political science, 'governance' has been first used in the context of international relations theories, in order to explain power relations and political structures that lack sovereignty and competence in order to enforce binding decisions. In contrast to traditional governmental systems that can be described with the term 'government', governance in this sense is understood as a response to the emergence of new forms of governing due to a shift from hierarchic, state-oriented steering, towards non-hierarchic forms of coordination between governmental and non-governmental actors across all levels, at times manifesting themselves in so-called policy networks (Mittag and Groll, 2010: 40–41). James N. Rosenau and Ernst Otto Czempiel's 1992 book *Governance without government* has been frequently quoted to illustrate 'a change in the meaning of government, referring to a new process of governing; or a changed condition of ordered rule; or the new method by which society is governed' (Rhodes, 1996: 652–653), and 'the development of governing styles in which boundaries between and within public and private sectors have become blurred [... and] governance mechanisms [...] do not rest on recourse to the authority and sanctions of government' (Stoker, 1998: 17).

Building upon these renderings of governance as forms of political steering and coordination within and beyond national boundaries, policy research has progressively dealt with parallel forms of collective regulation of societal issues, including self-regulation of private actors, co-regulation between public and private actors, and authoritative regulation through governmental actions (Mayntz, 2004: 66). Focusing on processes of implementation and steering, Mayntz's (2003: 28ff.) renewal of the 'Steuerungstheorie' (roughly translated as 'theory of steering') is related to the insight that, in the course of fulfilling complex societal tasks, the traditional strength of the state has decreased due to its loss of autonomy and the increasing collaboration with and between other private or public actors, manifested in the emergence of numerous collectively achieved sets of rules. This has been against the backdrop of seeing politics as management and interactions between various actors which contribute to the dissolution of clear distinctions between steering subjects and objects. In analytical terms, Mayntz and Scharpf (1995b: 16) equate governance with the aspect of regulation in complex structures, including external and self-regulation. They assume that regulation can occur in different ways of social action coordination, for example through reciprocal adjustments in the market or in interdependent actions, through agreements in negotiations, or through seeking consent in processes of communitisation.

All these translocations can also be described as follows. First, as a process of 'moving down' and locating tasks and competences to subordinate structures according to the principle of subsidiarity. Second, as a process of 'moving up' and shifting competences to the supranational level. And third, as a process of 'moving out' with the successive deregulation and privatisation of previously governmental sectors (Mittag and Groll, 2010: 41). In fact, Mayntz (2003: 34ff.) indicates that the concept of governance has to confront issues of transnationalism and globalisation, and she states that while, for assessing national governance structures, perspectives such as actor-oriented institutionalism have proven to be a useful and continually fruitful analytical approach, transnational or global governance nonetheless require further theorisation in order to try to systematise what Habermas has to referred as 'Die Neue Unübersichtlichkeit' ('the new obscurity'; 1985). With respect to specifying the added value of utilising the governance term in political science, Benz (2004: 19–21) argues that, since governance in general terms has been viewed as dealing with aspects of solving collective societal problems and thus political decision-making processes, a concentration on the three dimensions of polity, politics and policy seems feasible against the background of delineating governance from government perspectives.

The government perspective understands the state as an institution that differs from the market and society. The main focus is set on institutions of democracy, such as political parties and interest groups. Accordingly, political processes take place between these actors and are mainly decided upon by governmental organs. Policies, meanwhile, manifest themselves through legislation and the distribution of public services. The governance perspective in turn considers the state, the market, and social networks or communities all to be complementary forms of steering. Emphasis is therefore put on the steering and coordinating function of these institutional structures, finding its expression in conflicts, negotiations and adjustments between governmental and/or societal actors. In this sense, these agreements and compromises often lead to a mutual production of collective goods, as well as to modifications of institutional frameworks and network structures.

The term 'governance' and its applications are characterised by elasticity and vagueness, although the general implications are clear, i.e. regulation, steering and coordination. The reason for the impossibility of producing one definition stems from the fact that its scope of application is relatively broad. In political science, the concept is used to describe various focal points, such as new forms of governing in the European Union as well as developments taking place in public administration, in and between organisations, and in economic, labour and environmental policies. In addition, the content of the particular conceptualisation of governance depends on its respective geographic and historical context (Benz, 2004: 21).

Three specific applications shall be briefly exemplified: organisational, multilevel and global governance. Touching upon aspects of corporate governance, organisational governance focuses primarily on interactions between actors in an organisation (e.g. public authorities, associations or other organisations of the third sector), led by institutional control systems, and resulting in coordination of actions and resources in order to achieve strategic goals. It is therefore concerned with how steering processes can be optimised through institutional reforms, rather than through changing the actual processes. However, it does draw attention to the correlations between institutions and interactions and their respective procedural logic. As outlined by Schneider (2004: 174ff.), organisational governance concentrates on different steering and regulation arrangements within organisations and less between organisations, while it can be regarded as actor-oriented institutional cybernetics inasmuch as this approach analyses processes of system integration and adjustment, which are realised through highly differentiated institutional mechanisms. Referring to governance on the level of associations as 'associational governance', Schneider draws attention to Coleman's definition of corporate actors by illustrating the organisational dilemma actors confront when transferring resources and competences to the organisation. He therefore concludes that this approach can be useful in terms of analysing international non-governmental organisations (and in the context of global governance) with regard to identifying structural and institutional arrangements and their relevance in relation to the autonomy held by members or representatives, and the respective goal achievement probability.

Meanwhile, multi-level governance refers to steering and coordination processes across various levels. Besides drawing attention to aspects of territory and integration, the multi-level governance approach deals with institutional constellations deriving from intergovernmental and intra-governmental politics. As a consequence, the complexity of multi-level steering structures and their respective strategic implications under consideration of the institutional frameworks constitute central elements of analyses. In contrast to federalist approaches, multi-level governance enables the assessment of various multi-level structures (local, regional, national, supranational) through comparing problems of relations between (sub-)national and regional or international levels. As pointed out by Knodt and Große Hüttmann (2006: 227–229), multi-level governance does not see the role of states as a unit or collective actor at the forefront of analysis, but it rather focuses on utilising an actor-oriented approach by distinguishing between institutions (the state and the EU as a set of rules) and actors (individuals, groups or organisations acting within these institutions) in order to avoid abstract presumptions on national interests or preferences. Accordingly, this approach is based on the premise that competences are distributed across various levels (while rejecting the idea of viewing supranational entities merely as agents of principals), that supranational systems significantly reduce the competence of national governments (zero-sum game), and that political decision-making levels are interconnected (no clear separation between domestic and international politics).

Finally, global governance as an analytical perspective offers a conceptualisation of sector-specific forms of governing through international organisations or regimes, such as the World Trade Organisation and the World Anti-Doping

28 Theoretical perspectives and approaches

Agency. According to Rosenau (1995: 13), 'global governance is conceived to include systems of rule at all human activity – from the family to the international organization – in which the pursuit of goals through the exercise of control has transnational repercussions'. Forms of global governance are therefore shaped through coordination and cooperation between and amongst states and/or private actors. Considering the lack of a binding global institutional framework or hierarchy as provided by governments at the national level, global governance structures are therefore viewed as relatively labile. At the centre of analysis are thus the possibilities and difficulties of overcoming these coordination deficits through the development of a balance between processes of institutionalisation and informal self-regulation (Dingwerth and Pattberg, 2006). In a normative sense, global governance has also been used to refer to identifying new mechanisms of international politics in response to challenges deriving from globalisation (Behrens, 2004: 104ff.). Consequently, global governance frequently illustrates horizontal processes of self-coordination through the establishment of transnational advocacy and policy networks (Nölke, 2000).

To sum up, despite the variety of different analytical applications of the governance concept, Benz (2004: 25–27) identifies four core elements:

1. Governance signifies steering and coordination aimed at managing interdependencies between actors.
2. Steering and coordination are based on institutional control systems that are supposed to direct the actions of actors, while usually various combinations of control systems (market, hierarchy, negotiation) exist.
3. Governance also involves interaction patterns and modes of collective action that emerge in the context of institutions (networks, coalitions, reciprocal adjustments).
4. Processes of steering and coordination, as well as interaction patterns covered by the term 'governance', usually transcend organisational boundaries and in particular the boundaries of state and society, which have become fluid. Politics in this sense occurs through cooperation between governmental and non-governmental actors (or between actors within or outside of organisations).

Concluding that the analytical perspective of the governance term is not connected to a specific theory, Benz advocates the utilisation of institutionalist approaches with regard to assessing the impact of institutions, while the dynamics of institutional contexts, as well as the significance of institutional adjustments and structures of interaction, require an application of actor-oriented and historical institutionalist notions.

In relation to sport, Ronge (2006; 2010) has pointed out two thematic complexes of governance, namely the political science and sociological aspects of the governance term, and the societal sector of sport. Emphasising the traditionally non-commercial and non-governmental function of sport in society (at least in most western societies), in contrast to other policy areas, Ronge defines sport as being part of the so-called third sector since it is characterised by autonomous self-regulation, through clubs and federations which are mainly separate from governmental intervention apart from receiving public funding and financial support. Thus, the third sector and in particular sport could be regarded as a prototype of governance. The increasing commercialisation and professionalisation of the sport sector, accompanied by the interests and involvement of political actors such as the European Union, have led, according to Ronge, to transgressions of systemic boundaries resulting in a structural hybridity that, however, could only be solved by steering and coordination processes correlating with forms of governance.

Through identifying six perspectives in the context of sport policy and governance – sport as autonomous sector, as political symbol, as object of direct and indirect actions, as internal action, as political means, and in an international perspective – Tokarski *et al.* (2006: 6) have illustrated the complexity of the sport sector that can be covered by the governance term. In fact, Groll *et al.* (2010) have dealt with the emergence of global policy networks in sport as a specific form of governance with regard, for instance, to doping or racism issues. Concurringly, Steinbach (2006) has concluded that, through its wide perspective, governance is capable of pooling various theoretical approaches into a coherent (sport-)political science.

The role of AFC as 'skilful negotiator'

Based on the theoretical underpinnings, it can be presumed that the analytical dimension of the governance term has a considerable impact on the assessment of the role of AFC in Asian and global football politics. In this sense, governance is broadly understood as non-hierarchical forms of steering and coordination between governmental, economic and civil actors, framed by a variety of different institutional control systems and manifested in modes of interaction such as cooperation, coalitions, networks or reciprocal adjustments.

Specifically with regard to football, Sugden (2002) has developed the notion of network football drawing upon Castells' (1996) idea of a network society, and based on the assumption that the complexity of global football is marked by interrelations and conflicts between various actors such as federations, business companies, marketing agencies and political institutions while being dominated by global capitalist driving forces. Furthermore, Hindley (2002) has examined the utility of the concept of governance in relation to football, leading him to the conclusion that – while the term poses a great degree of elasticity – governance stresses problems of coordination across various agencies, organisations and political actors and therefore recognises the multi-layered diversity of interactions taking place.

Amara *et al.* (2005: 190–192) have drawn attention to 'the changing nature of sports governance' by arguing that globalisation processes have initiated 'a shift that is away from the government, or direct control, of sport in general and football in particular, to one of governance'. Contrasting between 'the days before multimillion sponsorship deals, media packages, [or] the European Union's Bosman ruling', when the traditional pyramid was characterised by 'FIFA as the ultimate authority in world soccer', and the contemporary setting, which is defined by a high degree of diffusion of competence and authority amongst players, clubs, federations, broadcasters, national governments, fans, media, sponsors and agents, the authors provide a definition of these new forms of steering and networking as 'systemic governance'.

Having been referred to by Holt (2009) in his study of UEFA and governance of football in Europe, Amara *et al.* (2005: 192) offer a useful starting point for studying the 'complex web of interrelationships between stakeholders in which different groups exert power in different ways and in different contexts by drawing on alliances with other stakeholders'.

Thus it is proposed that AFC could be defined as a 'skilful negotiator' in the local, regional and global structure of football, a system that is characterised by a complex web of interrelations containing a high degree of diffusion of competence and authority amongst players, clubs, federations, broadcasters, national governments, fans, media, sponsors and agents. AFC could be viewed as one element amongst many, with its structural and ideological sphere of influence varying and relying on functioning as a suitable coordinator and facilitator between all relevant actors.

As explained by Grix (2010: 160–170), 'the "governance debate" is [broadly speaking] between those that seek to emphasise the role of institutions and structures in the study of policy and those that attempt to focus attention on the beliefs and ideas of the actors within networks'. What he calls the 'governance narrative' has emerged from a shift of power from hierarchical top-down policy-delivery structures 'to one that is side-ways'. Differing from the Anglo-governance school – which focuses mostly on institutions and structures with regard to explaining changes from government to governance – and the interpretivist decentred approach – which views governance as implying bottom-up processes, while looking at the diversity of beliefs and ideas of (individual) political agents – Grix proposes a reconfiguration of both inasmuch as postulating an acceptance for the significance of 'structures and institutions, alongside a focus on the beliefs of actors', thereby allowing an adequate 'account for such "deviant" policy communities as sport and education that do not fit the original "governance narrative" '. Arguing that, in terms of ontology and epistemology, a dichotomous perspective (quantitative vs. qualitative, inductive vs. deductive, structure vs. agency) can hardly help understanding the real world, he embarks upon Max Weber's notion of '*erklärendes Verstehen*' ('interpretive understanding') in order to effectively provide 'a broad-brush, heuristic framework' and 'set the context for more fine-grained analysis of specific policies or policy delivery' by taking into account both actors' beliefs and ideas, and structures and institutions. He contends that 'social phenomena are complex and messy' and therefore ' "neat" theories are partial explanations at best, and can sometimes point us away from contextual, and often more interesting, information'. Drawing upon these observations, this study therefore views governance as a useful tool that 'offers an ideal scaffolding upon which to hang an argument', while looking at both ideas and beliefs and 'the role institutions and path dependency can play'. In fact, analyses of sport policies and politics 'via wider disciplinary approaches' could contribute to moving the academic study of sport 'from the margin to the mainstream'.

Notes

1 Woyke (2011: 212ff.) defines INGOs as associations of at least three non-governmental actors from at least three countries, operating internationally according to certain regulations. Mentioning that INGOs are subject to international private law, he also draws attention to so-called 'business international non-governmental organisations' (BINGOs), i.e. transnational corporations that are primarily interested in generating financial profits. In this sense he explicitly considers organisations such as the IOC, FIFA and UEFA, which have recently gained more and more political influence, to be INGOs. Nevertheless, whether confederations such as AFC should be termed international non-governmental or transnational organisations is subject to debate (Ritter, no year: 10–11). It very much depends on whether one prefers 'transnational' to 'international', while basically moving within a comparable spectrum of conceptualising similar phenomena, i.e. the emergence of new forms of politics including new actors that are captured through a wider analytical lens, rather than a state-centred and territory-based perspective (e.g. Colás, 2002).

For the purpose of this study, it seems feasible to bear in mind the fact that AFC is a confederation that includes 47 national member associations and operates across national borders within a specific region, while its headquarters is located in only one country in which AFC maintains the legal status of a non-profit entity. AFC can thus formally be considered a non-governmental organisation which, due to its activity as well as responsibility beyond the territorial borders of a state while consisting of members all related to defined geo-political borders, could therefore be understood as either a transnational organisation or an international non-governmental organisation. However, what makes the AFC different from an INGO such as Greenpeace is the fact that its actions are essentially restricted to the Asian continent. Also, it should be borne in mind that the commercial dimension of football and its respective tournaments sets major football governing bodies apart from other INGOs, albeit most are registered as non-profit entities.

2 Objecting to the concept of diminishing contrasts and increasing varieties, Amara *et al.* (2005: 204) argue 'that although there are global trends evident in all of the football systems reviewed, including, for example, the significance of merchandising, the role of TV rights, player influence and mobility', their comparative study on the governance of five professional football leagues:

> illustrates the converse of what Maguire and Elias refer to as increasing varieties and diminishing contrasts in relation to sport and culture respectively […] rather this case illustrates what might be more accurately described as diminishing varieties with increasing contrasts.

3 With regard to the political dimension of the globalisation of football, Giulianotti and Robertson (2009: 97ff.) utilise a two-fold approach consisting of a 'neo-mercantilist model of political-economic arrangements that define the interests of national and international institutions' but is being challenged by 'international governance as a further political-economic force'. In this sense, neo-mercantilism in football can be identified 'as operating principally among national football associations and national league systems, frequently in conjunction with the nation state itself.', e.g. the business-oriented

engagement of European national league systems and clubs in emerging football nations and regions leading to conflicts between foreign and local interests. As for international governing bodies such as UEFA and FIFA, the authors view them as having 'a dichotomous political function, relating on the one hand to neo-mercantilism and on the other hand to international governance'. To put it differently, 'in neo-mercantile terms, international governing body membership is comprised wholly of national football associations' and thus these 'bodies mirror national league systems in pursuing new spheres of economic influence'; on the other hand, 'international governing bodies differentiate themselves politically and ideologically by emphasising their governmental functions', namely their technical, legal and administrative competences, as well as their claim on 'moral custodianship of football'.

4 In the same book, Close/Askew's (2004: 243ff.) contribution draws attention to the problem of 'an unqualified orientalised interpretation of developments in East Asia – according to which an active, assertive Occident is the font of a cultural complex which is being successfully foisted upon a passive, supine Orient'. The authors express the opinion that 'football provides a case study which demonstrates that East Asia is actively engaged in a dialogue with the West rather than simply accepting the Western cultural account, or for that matter simply rejecting it.' Close/Askew reason that while 'the non-Western Other can be viewed as [a] helpless and passive victim of global trends – as exemplified by Edward Said's notion of Orientalism [...]. The Other can be (and in the case of football in East Asia) an active participant in the set of processes.' Therefore, they conclude that 'football as a cultural phenomenon [...], while originating in the West, is subject to glocalisation and playback' and its 'evolution is the product of mainly popular, bottom-up and democratic involvement'.

5 In comparison to 'old' institutionalism, which limited its view to forms of governmental organisation, neo-institutionalist perspectives have acknowledged the increasing significance of other institutions in political and social life. Accordingly, the term has been broadened with regard to including informal institutions such as social practices and routines (Morisse-Schilbach, 2006).

6 Hall and Taylor's (1996) enumeration begins with historical institutionalism. However, considering the fact that this approach bridges the gap between the other two in terms of utilising both a calculus and a cultural approach, it is here dealt with as third school of thought (see also Arnum, 1999).

7 Historical institutionalist works, however, have not been coherent. Instead, two streams have emerged with one focusing more on the rational behaviour of individuals and the other putting more emphasis on the cognitive dimension and how institutions influence individuals' identities. The rational version is characterised by a rather narrow understanding of institutions and looks at relatively short periods of time in which institutions are associated with generating path dependencies, and they are seen as structuring power in political power games. The sociological variant, on the other hand, follows a broader perspective by looking at longer periods of time in which institutions not only have a regulating but also a constitutive effect on actors' strategies and social identities (Morisse-Schilbach, 2006: 273–274).

References

Allison, L. (2005). 'Sport and globalisation.' In L. Allison (ed.), *Sport in the global society: The global politics of sport: The role of global institutions in sport* (pp. 1–4). London: Routledge.

Amara, M., Henry, I., Liang, J. and Uchiumi, K. (2005). 'The governance of professional soccer: Five case studies – Algeria, China, England, France and Japan.' *European Journal of Sport Science*, 5(4), 189–206.

Arnum, H. (1999). *Ideas and institutions in the European Union: The case of social regulation and its complex decision-making*. Copenhagen: Copenhagen Political Studies Press.
Bale, J. and Maguire, J. A. (1994). 'Sports labour migration in the global arena.' In J. Bale and J. Maguire (eds), *The global sports arena: Athletic talent migration in an interdependent world* (pp. 1–24). London: Cass.
Behrens, M. (2004). 'Global Governance.' In A. Benz (ed.), *Governance: Regieren in komplexen Regelsystemen* (pp. 103–124). Wiesbaden: VS Verlag für Sozialwissenschaften.
Benz, A. (2004). 'Governance: Modebegriff oder nützliches sozialwissenschaftliches Phänomen.' In A. Benz (ed.), *Governance: Regieren in komplexen Regelsystemen* (pp. 11–28). Wiesbaden: VS Verlag für Sozialwissenschaften.
Blum, S. and Schubert, K. (2011). *Politikfeldanalyse: Lehrbuch*. Wiesbaden: VS Verl. für Sozialwissenschaften.
Castells, M. (1996). *The rise of the network society*. Oxford: Blackwell.
Close, P. and Askew, D. (2004). 'Globalisation and football in east Asia.' In W. Manzenreiter and J. Horne (eds), *Football goes east: Business, culture and the people's game in China, Japan and South Korea* (pp. 243–256). London: Routledge.
Colás, A. (2002). *International civil society: Social movements in world politics*. Cambridge: Polity.
Coleman, J. S. (1974). *Power and the structure of society* (1st edition). New York: Norton.
Dingwerth, K. and Pattberg, P. (2006). 'Was ist Global Governance?' *Leviathan*, *34*(3), 377–399.
Eisenberg, C. (2006). 'FIFA 1975–2000: The business of a football development organisation.' *Historical Social Research*, *31*(1), 55–68.
Eisenberg, C. (ed.). (1997). *Fußball, soccer, calcio: Ein englischer Sport auf seinem Weg um die Welt*. München: Deutscher Taschenbuch-Verl.
Elias, N. (1994). *The civilizing process*. Oxford: Blackwell.
Elias, N. and Dunning, E. (2008). *Quest for excitement: Sport and leisure in the civilising process* (rev. ed.). Dublin: UCD Press.
Fozooni, B. (2004). 'Religion, politics and class: Conflict and contestation in the development of football in Iran.' *Soccer & Society*, *5*(3), 356–370.
Gammelsæter, H. and Senaux, B. (2011). 'Perspectives on the governance of football across Europe.' In H. Gammelsæter and B. Senaux (eds), *Routledge research in sport, culture and society, Vol. 3: The organisation and governance of top football across Europe: An institutional perspective* (pp. 1–16). London: Routledge.
Giddens, A. (1990). *The consequences of modernity*. Cambridge: Polity.
Giulianotti, R. and Robertson, R. (2007). 'Recovering the social: Globalization, football and transnationalism.' *Global Networks*, *2*(7), 166–186.
Giulianotti, R. and Robertson, R. (2009). *Globalization and football* (1st edition). Los Angeles: SAGE.
Grande, E. and Risse, T. (2000). 'Bridging the gap: Konzeptionelle Anforderungen an die politikwissenschaftliche Analyse von Globalisierungsprozessen.' *Zeitschrift für Internationale Beziehungen*, *7*(2), 235–266.
Grix, J. (2010). 'The "governance debate" and the study of sport policy.' *International Journal of Sport Policy*, *2*(2), 159–171.
Groll, M., Gütt, M. and Nölke, A. (2010). 'Globale Sportpolitiknetzwerke.' In W. Tokarski and K. Petry (eds), *Handbuch Sportpolitik* (pp. 142–157). Schorndorf: Hofmann.

34 Theoretical perspectives and approaches

Güldenpfennig, S. (1981). *Internationale Sportbeziehungen zwischen Entspannung und Konfrontation: Sport, Arbeit, Gesellschaft, Vol. 18*. Köln: Pahl-Rugenstein.

Güldenpfennig, S. (2011). 'Frieden durch Entwicklung und Entwicklung durch Frieden. Zu den Beitragsmöglichkeiten des Fußballs.' In K. Petry, M. Groll and W. Tokarski (eds), *Sport und internationale Entwicklungszusammenarbeit: Theorie- und Praxisfelder*. (Vol. 17, pp. 43–56). Köln: Sportverl. Strauß.

Habermas, J. (1985). 'The new obscurity: The crisis of the welfare state and the exhaustion of utopian energies.' Translated by Phillip Jacobs. *Philosophy & Social Criticism*, *11*(2), January 1986, pp. 1–18.

Hall, P. and Taylor, R. (1996). 'Political science and the three new institutionalisms.' *Political Studies*, (XLIV), 936–957.

Hargreaves, J. (1994). *Sporting females: Critical issues in the history and sociology of women's sports*. London: Routledge.

Hargreaves, J. (2002). 'Globalisation theory, global sport, and nations and nationalism.' In J. P. Sugden and A. Tomlinson (eds), *Power games: A critical sociology of sport*. London: Routledge.

Harnisch, S. (2011). 'Role theory: Operationalization of key concepts.' In S. Harnisch, C. Frank and H. Maull (eds), *Routledge advances in international relations and global politics: Role theory in international relations: Approaches and analyses* (pp. 7–15). New York: Routledge.

Harnisch, S., Frank, C. and Maull, H. (2011). 'Introduction.' In S. Harnisch, C. Frank and H. Maull (eds), *Routledge advances in international relations and global politics: Role theory in international relations: Approaches and analyses* (pp. 1–4). New York: Routledge.

Hayhurst, L. (2009). 'The power to shape policy: Charting sport for development and peace policy discourses.' *International Journal of Sport Policy*, *1*(2), 203–227.

Held, D. (1999). *Global transformations: Politics, economics and culture*. Cambridge: Polity Press.

Hindley, D. (2002). *An examination of the utility of the concept of governance in relation to the sports of swimming, football and cricket*. Unpublished dissertation, Loughborough University.

Hödl, G. (2002). 'Zur politischen Ökonomie des Fußballsports.' In M. Fanizadeh, G. Hödl and W. Manzenreiter (eds), *Journal für EntwicklungspolitikErgänzungsband, Vol. 11: Global players: Kultur, Ökonomie und Politik des Fußballs* (1st edition, pp. 13–35). Frankfurt am Main: Brandes & Apsel.

Holt, M. (2009). 'UEFA, governance, and the control of club competition in European football: A report funded by the FIFA João Havelange Research Scholarship.' Birkbeck Sport Business Centre report, Vol. 2, No. 1, January.

Hong, F. and Mangan, J. A. (eds). (2003). *Soccer, women, sexual liberation: Kicking off a new era*. London: Routledge.

Horne, J. and Manzenreiter, W. (2002). ,The World Cup and television football.' In J. Horne and W. Manzenreiter (eds), *Japan, Korea and the 2002 World Cup* (pp. 195–212). London: Routledge.

Horne, J. and Manzenreiter, W. (2004). 'Football, cuture and globalisation: Why professional football has been going east.' In W. Manzenreiter and J. Horne (eds), *Football goes east: Business, culture and the people's game in China, Japan and South Korea* (pp. 1–18). London: Routledge.

Houlihan, B. (1994). *Sport and international politics*. London: Harvester Wheatsheaf.

Houlihan, B. (1997). *Sport, policy and politics: A comparative analysis*. London: Routledge.

Houlihan, B. (2003). 'Politics, power, policy and sport.' In B. Houlihan (ed.), *Sport and society: A student introduction* (1st edition, pp. 28–48). Los Angeles: SAGE.

Houlihan, B. (2006). 'Politics and sport.' In J. Coakley and E. Dunning (eds), *Handbook of sports studies* (pp. 213–227). London: Sage Publications.

Houlihan, B. (2009). 'Mechanisms of international influence on domestic elite sport policy.' *International Journal of Sport Policy*, 1(1), 51–69.

Houlihan, B. (2010). 'International perspectives on sport structures and policy.' In W. Tokarski and K. Petry (eds), *Handbuch Sportpolitik* (pp. 48–62). Schorndorf: Hofmann.

Huntington, S. P. (1997). *The clash of civilizations and the remaking of world order* (1st edition). New York: Touchstone.

Kenis, P. and Schneider, V. (1987). 'The EC as an international corporate actor: Two case studies in economic diplomacy.' *European Journal of Poltical Research*, (15), 437–457.

Knodt, M. and Große Hüttmann, M. (2006). 'Der Multi-Level Governance Ansatz.' In H.-J. Bieling and M. Lerch (eds), *Theorien der europäischen Integration* (2nd edition, pp. 223–248). Wiesbaden: VS Verl. für Sozialwiss.

Lee, S. (2003). 'Moving the goalposts: The governance and political economy of world football.' In A. Budd and R. Levermore (eds), *Sport and international relations* (pp. 112–128). London: Frank Cass.

Lösche, P. (2010). 'Sportpolity, Sportpolitics und Sportpolicy als theoretische Annäherung an eine Sportpolitikwissenschaft.' In W. Tokarski and K. Petry (eds), *Handbuch Sportpolitik* (pp. 12–29). Schorndorf: Hofmann.

Maguire, J. (1994). 'Sport, identity politics, and globalization: Diminishing contrasts and increasing varieties.' *Sociology of Sport Journal*, 11, 398–427.

Maguire, J. (1999). *Global sport: Identities, societies, civilisations*. Cambridge: Polity Press.

Maguire, J. (2006). 'Sport and globalization.' In J. Coakley and E. Dunning (eds), *Handbook of sports studies* (pp. 356–369). London: Sage Publications.

Majumdar, B. and Bandyopadhyay, K. (2005). 'Race, nation and performance: Footballing nationalism in colonial India.' *Soccer & Society*, 6(2–3), 158–170.

Mangan, J. A. (2001). 'Soccer as moral training: Missionary intentions and imperial legacies.' In P. Dimeo and J. Mills (eds), *Sport in the global society: Soccer in south Asia: Empire, nation, diaspora* (pp. 41–56). London: Frank Cass Publishers.

Manzenreiter, W. and Horne, J. (2007). 'Playing the post-Fordist game in/to the Far East: The footballisation of China, Japan and South Korea.' *Soccer & Society*, 8(4), 561–577.

March, J. G. and Olsen, J. P. (1984). 'The new institutionalism: Organizational factors in political life.' *American Political Science Review*, (78), 734–749.

Mayntz, R. (2003). 'New challenges to governance theory.' In H. Bang (ed.), *Governance as social and political communication* (pp. 27–40). Manchester: Manchester University Press.

Mayntz, R. (2004). 'Governance im modernen Staat.' In A. Benz (ed.), *Governance: Regieren in komplexen Regelsystemen* (pp. 65–76). Wiesbaden: VS Verlag für Sozialwissenschaften.

Mayntz, R. and Scharpf, F. (1995a). 'Der Ansatz des akteurszentrierten Institutionalismus.' In R. Mayntz and F. Scharpf (eds), *Gesellschaftliche Selbstregelung und politische Steuerung* (pp. 39–72). Frankfurt am Main, New York: Campus Verlag.

Mayntz, R. and Scharpf, F. (1995b). 'Steuerung und Selbststeuerung in staatsnahen Sektoren.' In R. Mayntz and F. Scharpf (eds), *Gesellschaftliche Selbstregelung und politische Steuerung* (pp. 9–38). Frankfurt am Main, New York: Campus Verlag.

Meier, H. E. (2005). *Die politische Regulierung des Profifußballs* (1st edition). Köln: Sport und Buch Strauß.

Meyer, J. W., Boli, J., Thomas, G. M. and Ramirez, F. O. (1997). 'World society and the nation state.' *American Journal of Sociology*, *103*(1), 144–181.

Mittag, J. and Groll, M. (2010). 'Theoretische Ansätze zur Analyse europäischer und transnationaler Sportpoltik und Sportstrukturen.' In W. Tokarski and K. Petry (eds), *Handbuch Sportpolitik* (pp. 30–47). Schorndorf: Hofmann.

Morisse-Schilbach, M. (2006). 'Historischer Institutionalismus.' In H.-J. Bieling and M. Lerch (eds), *Theorien der europäischen Integration* (2nd edition, pp. 271–292). Wiesbaden: VS Verl. für Sozialwiss.

Nölke, A. (2000). 'Regieren in transnationalen Politiknetzwerken? Kritik postnationaler Governance-Konzepte aus der Perspektive einer transnationalen (Inter-) Organisationssoziologie.' *Zeitschrift für Internationale Beziehungen*, *7*(2), 331–358.

Nye, J. and Keohane, R. (1971). 'Transnational relations and world politics: An introduction.' *International Organization*, *25*(3), 329–349.

Oxford Dictionary Online. Definition of 'governance'. Retrieved from http://oxforddictionaries.com/definition/governance on 21 July 2011.

Powell, W. W. and DiMaggio, P. (1991). *The new institutionalism in organisational analysis*. Chicago: University of Chicago Press.

Rhodes, R. A. W. (1996). 'The new governance: Governing without government.' *Political Studies*, *44*(4), 652–667.

Ritter, M. (no year). *Der Wettbewerb um die Fußballweltmeisterschaft 2002*. Ruhr-Universität Bochum.

Robertson, R. (1990). 'Mapping the global condition: Globalization as the central concept.' *Theory, Culture and Society*, *7*(2–3), 15–30.

Robertson, R. (1998). *Globalization: Social theory and global culture*. London: Sage Publications.

Ronge, V. (2006). 'Governance: Begriff, Konzept unbd Anwendungsmöglichkeiten im Sport.' In W. Tokarski, K. Petry and B. Jesse (eds), *Veröffentlichungen der Deutschen Sporthochschule, Vol. 15: Sportpolitik. Theorie- und Praxisfelder von Governance im Sport* (pp. 9–20). Köln: Sportverl. Strauß.

Ronge, V. (2010). 'Das Governance-Konzept als Ansatz für die Sportpolitik.' In W. Tokarski and K. Petry (eds), *Handbuch Sportpolitik* (pp. 157–175). Schorndorf: Hofmann.

Rosenau, J. N. (1995). 'Governance in the twenty-first century.' *Global Governance*, *1*, 13–43.

Rütten, A. (1996). 'Zur Empirie der Macht: Soziologische Betrachtungen in einem unscheinbaren Politikfeld.' In G. Lüschen (ed.), *Sozialwissenschaften des Sports, Vol. 3: Sportpolitik. Sozialwissenschaftliche Analysen* (pp. 81–96). Stuttgart: Naglschmid.

Sage, G. (2006). 'Political economy and sport.' In J. Coakley and E. Dunning (eds), *Handbook of sports studies* (pp. 260–276). London: Sage Publications.

Scharpf, F. (2000). *Interaktionsformen: Akteurzentrierter Institutionalismus in der Politikforschung*. Opladen: Leske und Budrich.

Schmid, J. (1998). *Verbände: Interessenvermittlung und Interessenorganisationen; Lehr- und Arbeitsbuch*. München: Oldenbourg.

Schneider, V. (2004). 'Organizational Governance: Governance in Organisationen.' In A. Benz (ed.), *Governance: Regieren in komplexen Regelsystemen* (pp. 173–192). Wiesbaden: VS Verlag für Sozialwissenschaften.

Schneider, V. and Janning, F. (2006). *Politikfeldanalyse: Akteure Diskurse und Netzwerke in der öffentlichen Politik* (1st edition). Wiesbaden: VS, Verl. für Sozialwiss.

Schneider, V. and Werle, R. (1989). 'Vom Regime zum korporativen Akteur: Zur institutionellen Dynamik der Europäischen Gemeinschaft.' In B. Kohler-Koch (ed.), *Regime in den internationalen Beziehungen* (1st edition, pp. 409–434). Baden-Baden: Nomos-Verl.-Ges.

Slack, T. (2004). *The commercialisation of sport*. London, New York: Routledge.

Spitaler, G. (2009). 'Politikwissenschaft und Sport.' In M. Marschik, R. Muellner, O. Penz and G. Spitaler (eds), *Sport Studies: Eine sozial- und kulturwissenschaftliche Einführung* (pp. 273–275). Wien: Facultas.

Steinbach, D. (2006). 'Das Governancekonzept als innovativer Ansatz für die Sportpolitik und Sportpolitikforschung.' In W. Tokarski, K. Petry and B. Jesse (eds), *Veröffentlichungen der Deutschen Sporthochschule, Vol. 15: Sportpolitik: Theorie- und Praxisfelder von Governance im Sport* (pp. 21–30). Köln: Sportverl. Strauß.

Stoker, G. (1998). 'Governance as theory: Five propositions.' *International Social Science Journal*, 50(155), 17–28.

Sugden, J. (2002). 'Network football.' In J. P. Sugden and A. Tomlinson (eds), *Power games: A critical sociology of sport* (pp. 61–80). London: Routledge.

Sugden, J. and Tomlinson, A. (1998). 'Power and resistance in the government of world football: Theorizing FIFA's transnational impact.' *Journal of Sport and Social Issues*, 22(3), 299–316.

Sugden, J. and Tomlinson, A. (2005). 'Not for the good of the game.' In L. Allison (ed.), *Sport in the global society: The global politics of sport: The role of global institutions in sport* (pp. 26–45). London: Routledge.

Tokarski, W., Petry, K. and Jesse, B. (2006). 'Vorwort.' In W. Tokarski, K. Petry and B. Jesse (eds), *Veröffentlichungen der Deutschen Sporthochschule, Vol. 15: Sportpolitik: Theorie- und Praxisfelder von Governance im Sport* (pp. 5–6). Köln: Sportverl. Strauß.

Weaver, R. K. and Rockman, B. A. (1993). *Do institutions matter? Government capabilities in the United States and abroad*. Washington, D.C.: The Brookings Institution.

Whannel, G. (2008). *Culture, politics and sport*. New York: Routledge.

Williamson, O. E. (1979). 'Transaction-cost economics: The governance of contractual relations.' *Journal of Law and Economics*, 22(2), 233–261.

Woyke, W. (2011). *Handwörterbuch internationale Politik* (12th edition). Opladen: Verlag Barbara Budrich.

Woyke, W. and Delschen, A. (eds). (2006). *Sport und Politik: Eine Einführung*. Schwalbach am Taunus: Wochenschau-Verl.

3 Football in Asia
Developments and challenges

Asia, as the largest continent with the highest population in the world, is characterised by great geographical, social, cultural and economic diversity. Accommodating a third of the world's population, it comprises three main cultural areas, namely the Islamic, the Hindu-Buddhist and the Sinic. Yet almost all Asian countries have undergone similar historical developments and transformations, from political independence to modernity, 'through a triadic process of cultural continuity, assimilation of contemporary ideas and resistance to imperial power' (Hong, 2002: 401–402).

In this political and cultural context, modern sport has constituted an eminent factor in reassessing political structures and developing a regional identity. Not least, the Asian Games have since their inauguration in 1951 contributed significantly to this development. Having come out of the shells of colonialism, Asia has experienced a rapid transformation ever since the 1980s. Being 'a centre of modernity', globalisation has affected Asia and its sports. Despite the political, cultural, ideological, economic and social diversity, similar developments and challenges have occurred with regard to changes in sport systems, the roles of the media and sponsorship, and the significant increase of sport-industrial matters. However, the question remains how Asia and its countries perceive, handle and respond to these constellations as well as how they manifest themselves, particularly in relation to the most popular sport of all, namely football.

At the AFC Congress in 2010, then-President Mohamed Bin Hammam spoke to the delegates and mentioned the importance of capturing the history of football in Asia in order to understand and appreciate today's achievements and challenges. He began by referring to a finding according to which 'the great people in China had invented this beautiful game five thousand years ago in the city of Zibo and we've introduced it to the world via the old Egyptians, Greeks and Romans'. Indicating that the modern game of football, however, originates from Britain, he continued by mentioning that it then 'was reintroduced to our continent by the British and the first match ever recorded took place in India in 1854 – 100 years before the establishment of the AFC'. In consideration of the history of Indian football, he also mentioned the founding of the first Asian club Mohun Bagan in 1889. (Quoted in Asian Football Confederation, 2010: 26–27.)

This pattern of historisation has been increasingly drawn upon in recent years due to expansion into Asian football markets accelerated by clubs, sponsors and federations alike. Official histories, such as those produced by FIFA and AFC, therefore acknowledge the ancient football traditions of the Asian region, namely *cuju*, which was an ancient ball-kicking game practised in China in the time between the second and third centuries BC, as well as *kemari*, a Japanese game played ever since the 12th century AD (FIFA, no year; Asian Football Confederation, 2004: 14–18). However, 'most academic histories continue to prioritize direct evidence to emphasize the British origins of modern football' (Giulianotti and Robertson, 2009: 5). Accordingly, Goldblatt (2006: 9) remarks that: 'What has ultimately determined which game modernity plays is not who played the game earliest, nor even who kept the ball on the ground for the longest, but who played it at the moment of modernization.' Furthermore, the fact that modern football, otherwise known as association football, was regulated and institutionalised by officials long after most ancient games had disappeared, puts a continuity thesis into question, albeit traditional ball games had also existed in Britain. The establishment of binding rules and the founding of the English Football Association in 1863 should hence be considered as the actual birth of modern football, thereby laying the basis for its growth and reproduction throughout the world (Eisenberg, 2004: 7ff.). The universal quality of the regulations and principles provided grounds for a cultural transfer which, however, required one of the major technical inventions made in the era of industrialisation, the steamboat. Ever since the late nineteenth century, the steamboat contributed to a significant increase in intercontinental transportation, thus intensifying overseas migration between Britain and other continents. Wealthy tourists, businessmen, workers and soldiers soon turned out to be decisive in spreading the game and its popularity (Eisenberg *et al.*, 2004: 38–44).

The time between the 1870s and mid-1920s, entitled the 'take-off phase' by Giulianotti and Robertson (2009: 7–14), covers the period in which 'football became embedded within the cultures of Europe and South America, and in "Europeanized" parts of Africa, Asia and North America'. While rugby and cricket were popularised through colonial mechanisms, football expanded mainly 'through a "trading ecumene" via business and industrial routes and in relatively informal social ways'. It was everyday social relations, such as British maritime and industrial workers playing football matches in ports of South America or Hong Kong, that 'contributed massively to the incubation of football within these settings, as local people watched the British play and, in emulation, were inspired to found their own clubs'. The reception of the game, however, did not occur without obstacles. In colonial environments, football matches were instrumentalised for nationalist purposes and as a means of expressing resistance against occupation, such as in India where local victories over British teams were interpreted as revealing the emancipatory potential in terms of gaining independence from the Empire.

In contrast to Africa, Asia met European colonialism with a profoundly higher degree of resistance, both in military and cultural terms. Japan, for instance, could not be conquered and India, despite the British occupation, largely maintained its culture and tradition. Another factor was the presence of Americans in some east Asian countries where missionaries, tradesmen and – such as in Japan or Korea – soldiers popularised baseball and made it not only a 'significant competitor to football' but in fact 'the most popular sport for most of the twentieth century'. It also has to be borne in mind that, besides Asia having been a region intensively affected by wars and inter-state cross-border conflicts throughout the twentieth century, and thus seriously destabilised in terms of socio-economic progress, not all people were exposed to the above-mentioned social contacts since in many Asian countries the majority of the population lived in rural areas, whereas football 'is a game of the city' (Goldblatt, 2006: 538–540).

The subsequent development of football on the Asian continent is all the more fascinating considering the fact that it has become 'the people's game', reaching into almost all parts of Asia and experiencing matches with sold-out stadiums counting over 100,000 spectators in India and 150,000 in Indonesia (Murray, 1995: 139). Less than two decades ago, Asian football was represented by just two teams at the World Cup 1990 in Italy, professional football leagues were virtually non-existent or of hardly any sporting and commercial importance, and the discrepancy between the Asian periphery and the European and South American centres of football grandeur seemed insurmountable, despite the efforts that had been made ever since the 1960s and 70s with regard to developing and professionalising the game. However, only 12 years on from World Cup 1990, Asia hosted its first World Cup tournament, with South Korea advancing to the semi-finals and millions of fans enjoying the spectacle. In fact, 'when the world sits down to watch the World Cup Final more than half the audience is now in Asia' (Goldblatt, 2006: 831).

Bearing these circumstances in mind, the following section takes a look at the rise of football and its tensions, popularisation, commercialisation and professionalisation in the regions of east, southeast, south, central as well as west Asia. The selective overview contains exemplary descriptions of Hong Kong, South Korea, Japan, the World Cup 2002, China, Singapore, Malaysia, India, Iran, Saudi Arabia, and Qatar (World Cup 2022).

Hong Kong

The first football governing body in east Asia was neither established in Korea nor Japan – today's powerhouses in the region – but in the former British colony Hong Kong. The Hong Kong Football Association was founded in 1914, and it paved the way for prestigious tournaments such as the Hong Kong Shield and the inauguration of the first professional football league in the region in 1968. Inspired by the prior launch of the North American Soccer League, it could not, however, generate a similar significance and thus suffered from several structural failures while not being capable of attracting major top players. Instead, most imports turned out to be out-dated Europeans or young talents. Nonetheless, this initial attempt contributed to the overall popularity of football in the region and helped set the basis for other professional leagues to come.

In terms of football administration and politics, meanwhile, Hong Kong belonged to the pioneering regions. As one of the founding members of AFC, Hong Kong used to host its secretariat and was represented by HKFA President Henry Fok, thereby attaining influence at the regional level. While the country hosted the first Asian Cup in 1956 and reached the third round, the national team could not achieve any considerable victories at international level. The poor performances have not contributed to the sustainability of the domestic league and its popularity. The interest in the league reached a low in the 2000/01 season when a total of only 55,935 spectators went to attend all 81 matches. The crisis required the HKFA to invite teams from China to maintain a sufficient number of teams, while a restructuring of the HKFA's decision-making body was implemented in order to make processes more effective. While the heyday of football in Hong Kong during the late 1970s and early 1980s could not be revived, and despite the affiliation with China in 1997, Hong Kong still has its own football governing body, a league consisting of 13 teams, as well as a representative national team (East Asian Football Federation, 2005: 62; 2011; Murray, 1995: 140; Football Asia, 2002a: 150).

South Korea

The evolution of association football in Korea was severely inhibited by the geopolitical situation the country found itself in from the beginning of the twentieth century until the 1950s. Japan occupied Korea, due to which all major football activities suffered a lot. In fact, since the country was not sovereign any more, Korean players, if interested in competing at the international level, had to play for the Japanese national team. Although Korean city teams were successful and even won some international club competitions, the Japanese decided to nominate only two Koreans for the Olympic squad in 1936. This development eventually culminated in the abolition of the national association in 1942. A change came, however, with the end of the Second World War and Korea gaining independence from its occupiers. Shortly after the war, the Korea Football Association was reinstated in 1945, 17 years after its initial foundation. This association even endured the Korean War and the separation of the country into North and South Korea, being thereafter solely responsible for the southern part. The KFA has also been affiliated with FIFA since 1948.

After the South Korean team had succeeded in qualifying for the World Cup 1954 in Switzerland, further tournaments could not be reached until 1986. Meanwhile, the Asian Cup was lifted in 1956 and 1960. Due to the shortage of success at the international level, the decision was made to foster the local football scene through the establishment of a professional league, the Korean Super League, in 1983. At first, the league consisted of merely two professional and three amateur teams, which competed for the title in a rather unorthodox format without any home and away matches, but instead the teams had to travel the country and play in various cities. Despite its relatively low popularity, the league helped get the national team back onto successful international tracks, qualifying for all World Cups since 1986 (Lee, 2002: 78–79; Korea Football Association, 2011).

In the context of an economically rising South Korea ever since the Park dictatorship, the government focused on using sports as a means of promoting the nation's wealth, hence the successful bid for the Seoul Olympics and the establishment of professional baseball and football leagues. In the post-authoritarian era, the KFA has in fact become 'an agent of government in the management of government programs' (Chung Hongik, 2004: 127). In exchange, the KFA receives financial, administrative and financial support, especially in relation to the national teams, while at the grassroots level football is still under-funded and badly managed. Moreover, the approach means that KFA officials and in particular the President have a political rather than a sports background. The long-time KFA President and former FIFA Executive Committee member Chung Mong-Joon, for instance, is an entrepreneur and a politician who ran for the South Korean Presidency in 2002 (Manzenreiter and Horne, 2007: 569).

The number of clubs and teams in the Super League, meanwhile, grew so that the league expanded and was revamped in 1996, henceforth going by the name of K-League. In the build-up to World Cup 2002, attendance figures rose towards the end of the 1990s and more clubs joined the league, so that by the turn of the millennium 12 teams with 450 registered players competed for the title. With the season ranging from spring until autumn and covering a total of 200 matches, the league attracted 2.3 million spectators in 2001 (Chung Hongik, 2004: 122). In 2005, the league consisted of 14 and in 2012 of 16 teams. By 2005, the average spectator attendance had increased to 10,000 per game (Höft et al., 2005: 181). The KFA advanced its organisational structure through introducing more departments specialised in developing the domestic game. Furthermore, according to a KFA employee (Hoon Han, 2011), the governing body:

> built three more football centres in central, eastern, and southern part of Korea. In addition, 20 football parks have been built around the country, which give youngsters more chances to be involved in football under the better circumstances [... while ...] the long-term vision in connection with the strong organisational structure has to be implemented for the preparation not only for the performances at the moment but also for the 10 or 20 years ahead.

Two significant aspects of the K-League are the franchising system, which like the North American model allocates a club to a geographic location, and the involvement of commercial entities that own and sponsor most of the teams. This explains the names of teams such Jeonbuk Hyundai Motors and Suwon Samsung Bluewings (Horne and Manzenreiter, 2007: 271). With the risk of having to modify names when sponsors change, clubs and especially their fans have had problems creating a sustainable identity and tradition. Occasionally, some clubs moved from the cities and urban areas to other locations, thereby losing a considerable amount of their constituency. The reason for moving the clubs was due to their concentration in the greater Seoul area; accordingly, the decision was made to resettle them to other local communities. Towards the end of the 1990s, attempts were made to make the clubs more independent from the companies through encouraging the removal of the sponsors from the club names. These frequent modifications though, and production 'of a completely artificial team location, without a basis in tradition, history or culture', also had commercial implications due to 'the lack of passion' and identity (Ravenel and Durand, 2004: 36).

In comparison to its male equivalent, women's football in Korea has hardly received a similar amount of public attention and thus its development has been rather slow and marginal. Despite the fact that games between females have a 40-year history, with the First National Women's Football Games dating back to 1974, traditional gender relations and social structures led most football authorities to the conclusion that women were not suitable for playing the game. Even FIFA's encouragement during the mid-1980s to institutionalise women's football and create a national team ended in failure. Only due to the public attention generated by the World Cup 1999, hosted in the US, did South Korean officials recognise the potential of improving the standard of play and therefore the structures of the sport in their country. Subsequently, the KFA began making investments and established the Women's Football Federation in 2001. In spite of initial victories that the national team achieved and a temporary wave of popularity, women's football experienced a rapid decline only a year later. Due to infrastructural deficits and low-quality development schemes at both the grassroots and elite levels, South Korean women's football has henceforth not managed to overcome sexist stereotypes and mainstream the sport into South Korean culture (Koh, 2003: 72ff.).

Recently, a major corruption and match-fixing scandal, which revealed some of the failures and deficits of the system, struck South Korean football. A number of players were accused of manipulating games and 47 former and current players received a lifetime ban. Although punishments were handed out and players' salaries were increased as a preventative measure, the KFA has announced that further measures may be necessary in order to prevent criminals from infiltrating the game (Soccerex, 2011). Furthermore, the K-League has considered introducing drastic reforms by introducing a promotion and relegation system aimed at raising the attractiveness of the league, thereby increasing its commercial value, in order to overcome the consequences of the crisis (Somerford and Kim, 2011).

Japan

Ten years after South Korea had installed its professional league, the Japanese caught up by founding the J-League in 1993. It had 'the major objectives of furnishing advanced sports facilities, building better training systems for players and [...] also providing opportunities for as many people as possible across the country to experience the joys of sport' (J-League, 2011).

44 *Developments and challenges*

After football in Japan had been facilitated by the migration of British workers, football matches at a competitive level mostly took place at schools, universities and companies, while a precursor to today's league, named the Japan Soccer League, was founded for the company teams in 1965 (Giulianotti and Robertson, 2007: 180). Although ever since 1986 it was possible for the companies to award players professional status, this situation meant that professional football careers and adequate salaries were only accessible for employees of particular companies (Höft *et al.*, 2005: 178). The Japanese Football Association, which was founded in 1921 and admitted to FIFA eight years later, was determined to change this situation by nurturing a football culture embedded in a local club structure, thereby creating access for more players and coaches (Manzenreiter, 2004a: 293). In fact, former J-League Chairman and Chief Executive Saburo Kawabuchi said that they 'were determined to build a new sports culture for Japan' (quoted in Sugden and Tomlinson, 1998: 171).

In contrast to the South Korean system, governmental actors have been less concerned with sports, thus giving the decision-makers relative flexibility for organising the new league, while having to deal with a context in which baseball was a well-established part of culture and the national sports industry. Also, the fact that the Japanese economy had experienced transformations throughout the 1980s, requiring new service-oriented and post-industrial approaches, defined the setting in which the J-League had to be established and popularised. At the very beginning, the J-League administrators tried to avoid structural mistakes that had been made in South Korea and decided that none of the ten founding clubs were to come from Tokyo. The intention was to establish a sustainable fan culture in the suburban and rural areas as a basis for improving the quality of Japanese football in the long run. Several merchandising products were designed, including new logos and mascots, former international star players such as Gary Lineker and Zico were signed, and sponsors such as Mizuno were urged to initiate large-scale marketing campaigns. Statutes were brought in according to which clubs were prohibited from incorporating the sponsors' names into their club name because, as in South Korea, most of the teams originated from corporate backgrounds. Nonetheless, this did not prevent fans and spectators from knowing exactly which company supported which team. The respective corporations, meanwhile, were encouraged to run the clubs professionally and to establish special departments within the company only responsible for dealing with club-related matters (Ichiro, 2004: 44ff.).

The J-League can be considered a success story by Asian standards, not only indicated by the impact it had on performances at the international level with the Japanese team qualifying for the World Cup 1998, but also as acknowledged by AFC and FIFA, who both praised the professional and well-structured management of football development in Japan (Sugden and Tomlinson, 1998: 173). J-League and club officials had been highly influential in shaping the Japanese fan culture, synthesising fan practices from European and South American stadiums with traditional Japanese spectator culture (Giulianotti and Robertson, 2007: 181). Indeed, the first three years saw an average of 20,000 spectators per match coming to the stadiums, but due to the economic crisis in the late 1990s ticket sale numbers decreased rapidly, leading to the recognition by the companies that an entire professionalisation of the club section may be wiser. Hence, they began with recruiting competent sport managers and educating their personnel according to the requirements of professional sport management concepts (Horne and Manzenreiter, 2007: 271).

Moreover, the JFA contrasted the J-League with the rather old-fashioned baseball scene by focusing on attracting female fans to football. The media strategy therefore concentrated on presenting players and their general appearance in a fashionable way. Respective magazines and newspapers reported on players' haircuts, their favourite clothes and their entire lifestyle, while sport-related matters were superficially touched upon (Höft et al., 2005: 180). However, women's football as an activity on the pitch has not been particularly encouraged or appreciated. Despite the national team's success and the instalment of the first Asian professional women's league, football and the representation of Japan at the international level are still largely masculine domains. Gender relations, however, are not static and football therefore can be regarded 'as one battle field for the reconstruction of the gender order in Japanese society.' (Manzenreiter, 2008: 255)

In comparison to the South Korean and Chinese leagues, the J-League underwent a relatively steady and positive development. The average number of spectators stabilised after it had reached its lows in the 1990s, the J-League modified its official slogan from 'A 100 year project' to 'A happier nation through sport', squad sizes were limited and the number of and salaries of foreign players were reduced, authentic fan cultures emerged based on the fact that the expectations towards the league had become more realistic in comparison to the initial euphoria of the first few years, and the World Cup 2002 left a footprint and raised awareness of Japanese football at the global level (Horne, 1999: 224–225). A J-League employee (2011) commented on these aspects as follows:

> It was phenomena as J-League was the first every professional football league in Japan. Also, it was very attractive, as the way to professional was different from Professional baseball, which already has been the professional sports. [...] World Cup gave many world-standard stadium in Japan, and all the stadium is now used as J-League club's home stadium [...] everybody who was not familiar with football had many chance to see world famous players and happy supporters on streets and TVs. World Cup definitely gave us a chance to overcome baseball, which is the most popular sports in Japan. [...] Mr. Kawabuchi's passion for professionalization was the biggest reason. Our football development is definitely based on the foundation and expansion of J-League.

With the 'JFA declaration 2005', the national governing body issued a document revealing its roadmap and vision for the twenty-first century. The mid-term objective to be reached by 2015 is to expand the overall Japanese football membership to five million while making the national team one of the best ten teams in the world. And in the long run by 2050 the JFA aims at not only enriching Japanese society and expanding people's minds through football, but also at hosting the World Cup once again in Japan, as well as celebrating the national team as winner of the most prestigious tournament. In order to realise these ambitious objectives, the JFA has collaborated actively with its 47 sub-associations by conducting extensive analyses aimed at a fundamental reforming process, which aims to lead to the ideal scenario in 'which future generations may say: "We have come this far thanks to the Declaration"' (Japan Football Association, 2005).

World Cup 2002

In 1986, AFC informed FIFA that it would appreciate hosting the World Cup in 1998 or 2002 (Asian Football Confederation, 1986: 6). When the prospect of being able to host the tournament in Asia first seemed realistic, it was initially Japan that engaged in creating a campaign. A couple of years later, South Korea joined and at this stage of the process both countries were still determined to host the competition by themselves. It soon became clear that Japan and South Korea had very different understandings of hosting the tournament and hence developed different bidding strategies. Even with regard to the process of appointing personnel to the bidding committees, they deviated from each other. While South Korea focused on picking primarily people from the political scene, Japan drew upon personalities from the sports sector. After getting to know the South Korean way, however, Japan decided to appoint former Prime Minister Miyazawa Kiichi as Honourable Chairman of their committee.

Further differences existed in relation to how the committees were supported by the respective governments. The South Korean political elite showed immediate and unlimited endorsement for the bid, and the President not only met with high-ranking FIFA officials and made the case for South Korea during his trips to Europe, but he even wore sport clothes which advertised the bid while going jogging. The Japanese committee by contrast was almost left alone in terms of political support. The final approval was issued only a week before the bidding deadline ended in February 1995. Accordingly, it can be assumed that the bid for the tournament had very a different degree of prestige for both countries. The South Korean case indicates that it was considered a major national project, while the Japanese approach suggests the assumption that football and its most important tournament was mostly of sporting significance. As a matter of fact, the South Korean perspective reinforced the political dimension of hosting the tournament by referring to its potential for contributing to the geo-political stability of the region and bringing North and South Korea closer to each other. Moreover, regular references were made to the historical relationship with its competitor Japan by outlining the dreadful incidents that had occurred during the Japanese occupation in the early twentieth century. The Japanese, however, neither reacted to these provocations nor produced any other political aspirations (Butler, 2002: 44–46).

Developments and challenges 47

The Japanese perceived the competition with South Korea in a more sportive way. Former professional player Okano Shun'ichiro commented on the issue by pointing to the unfortunate consequence of having a winner and a loser, but warned that the final result should not intensify hatred and anger between the nations. Despite receiving general praise for this standpoint, it seemed that South Korea's strategy was more effective and it was also backed by the majority of the European member associations, while the Japanese had FIFA President João Havelange and his supporters on their side. As a matter of fact, at the beginning of the 1990s Japan was the first of the two countries engaging in building up a campaign, and initially it seemed undisputed that with Havelange's support the decision was basically only waiting to be carved in stone (Ritter, no year: 67). The final decision to host the tournament in both countries, however, was based on an idea put forward by AFC in March 1996. UEFA had then become fond of the suggestion and henceforth supported this alternative. Finally, a FIFA delegation investigated the issue and opined that both bids were not too dissimilar and that a bilateral holding seemed sensible (Butler, 2002: 49).

The new solution did not only rest upon the recognition that the bidding process had become inharmonious, but also that both countries had made remarkable financial investments which in the event of a rejection would have led to both a significant financial detriment and a loss of face. With the decision of co-hosting the World Cup in Japan and South Korea, FIFA had prevented itself from being accused of being responsible for the defeat of one of the countries. Furthermore, AFC benefited from the decision, first because it had solved the conflict between its constituent member associations, and second because now one more Asian team would participate in the tournament due to the fact that both host countries were to qualify directly (Dunkel, 2009: 42).

At the very beginning of the bidding process, the leading Japanese advertisement and marketing agency Dentsu published several studies relating to the expected economic benefit of the World Cup. Interestingly, Dentsu was very optimistic about it, not only due to its obligations towards its partners but also since it was responsible for all merchandising products in relation to the World Cup, as well as due to the fact that it held the broadcasting rights for the matches of the Japanese team. Accordingly, Dentsu predicted an economic net effect of up to 0.6 per cent of the GDP. On the South Korean side, similarly optimistic forecasts were made: the Korean Development Institute estimated an employment creation of 350,000 and an industrial output of US$8.8 billion. After the South Korean team had successfully played its first few matches, the Hyundai Research Institute presented an updated estimate of US$11.5 billion for the victory over Poland alone. The overall benefit after South Korea had lost to Germany in the semi-finals was then predicted to have increased up to US$77.8 billion (Manzenreiter, 2004b: 68–69).

The actual economic effects, however, were difficult to calculate, especially in the South Korean case. Generally speaking, what harmed the tournament's financial outcome most was the troublesome sale of tickets, especially abroad. According to independent research carried out by Ernst and Young, FIFA had to make a one-time payment of US$100 million in order to compensate for the losses. Also, the Japanese World Cup Organising Committee (JAWOC) performed its accounting with a certain degree of creativity, thereby reducing some of the expense matters through omitting the costs of the bidding campaign, as well as the salaries and overhead expenses of JAWOC itself. Meanwhile, the highest expenses had to be covered by the local communities which had invested huge sums, not only into stadiums and infrastructure but also into the bidding campaign and the costs for the JAWOC. In fact, some of the credits that had been needed for building the stadiums are not scheduled to be completely repaid until the 2030s. The supposition according to which the expenses could be compensated through national league matches and other events was false. In 2003, the total loss amounted to Yen 2.5 billion (Manzenreiter, 2003: 228–233).

In socio-cultural terms, meanwhile, the World Cup has turned into a national landmark for both countries, which has been incorporated into the collective memories of the respective populations (Manzenreiter, 2003: 226). In South Korea, most people perceived the period of the World Cup as one of the best times in their lives. Some even compared the significance of the event with the day when the Japanese occupation ended in 1945. In particular, the positive performance of the team contributed to the previously unimaginable enthusiasm and excitement amongst the fans. The fact that hundreds of thousands gathered in public places, wore red shirts and celebrated together created a sense of national unity, pride and new confidence. This experience was emphasised by players, officials, politicians, journalists and fans alike as unique and unprecedented. It was, however, not an experience the South Korean nation made in interaction with other nations, but rather an exclusive South Korean phenomenon in which the people could experience themselves as the South Korean nation. This has to be understood against the background of how the South Koreans perceived their own history, the colonial era, the Korean War, the separation of the country and the dictatorship. In this sense, the World Cup had a certain catharsis effect. As a matter of fact, the way the younger generation employed national symbols was new. While Japan aimed at internationalising itself through the tournament and synthesising the local and the global, South Koreans sought the opportunity of utilising the event and the success of its team as a means of presenting its nation, especially in relation to Japan, as equal and on par with other countries.

More impressive was the reaction the Japanese showed, despite the fact that their team had been eliminated at an early stage. Instead of taking the side of whoever played against South Korea, they supported the South Korean team, thereby creating a very positive experience with its historical 'arch enemy', fostering intercultural dialogue and generating a sense of unity. Therefore, it can be concluded that the bilaterally hosted World Cup, which had been put on rather involuntarily, helped with achieving some of the originally formulated objectives and expectations more effectively than would have been the case if one of the countries had been the sole venue (Dunkel, 2009: 99ff., 153ff., 182).

China

Despite the frequent reference to China as the origin of football, where ancient ball games had been played 2,300 years ago, the modern game of association football was not introduced until the late nineteenth century, when British migrants facilitated the sport in the major commercial locations such as Shanghai, Beijing and Hong Kong (*People's Daily*, 2004). After some clubs had been founded, such as the Shanghai Athletic Club in 1887, followed by the creation of a Shanghai football association in 1910, football underwent a relative decline before reflourishing in the early twentieth century, leading to the establishment of the South China Athletic Association in 1904, which marked the beginning of an era during which the SCAA won a prestigious regional tournament called the Hong Kong Shield in 1929. At the time, teams consisted of players with a middle-class background or students who played for teams owned by wealthy businessmen. It was common that the teams were composed upon ethnic background, often leading to conflicts on and off the pitch. By the 1930s, however, previous all-European teams included Chinese players, so that locals also became involved in administrative matters. As a result, football was institutionalised and the Chinese Football Association was founded in 1924 and admitted to FIFA in 1931, while a national team was put together in order to compete at the 1936 Olympics in Berlin (Murray, 1995: 149–151).

In light of the Second World War, the Japanese invasion and the emergence of the socialist revolution, China underwent a period of political, economic and social transformation which affected the development of football. With the declaration of the People's Republic in 1949, sports in general and football in particular were regarded as symbolising modernity and as a platform for portraying the new China. In 1952 the entire sport system was centralised and put under the control of the National Sports Commission (NSC), which was henceforth responsible for formulating and implementing sport policies and programmes in close liaison with the governmental authorities. In the context of the Cold War, contacts were established with other communist countries in relation to fostering sport development, especially of high performance and elite sports. While at first reaping the fruits of centralised football training and development methods – complemented by the establishment of a national league system in 1956 with a considerable number of specialised football teams at municipal and provincial levels, thus making China a footballing powerhouse in east Asia – the country then experienced a dramatic economic downturn (Jinxia and Mangan, 2002: 79–82). In the course of the Cultural Revolution, initiated in 1966, the Chinese sport system suffered a severe setback, seeing its star athletes sent to the fields, administrators imprisoned, and sports facilities closed down. By 1958 international sport contacts had been cancelled in protest against the IOC's 'Two Chinas' policy, followed by withdrawals from nine other organisations including FIFA.

This period of isolation dissolved in the early 1970s with 'ping pong' diplomacy, Mao Tse-Tung's death in 1976 and the rise of Deng Xiaoping, who began initiating a new economic and international outlook. In this context the sport system was reformed, with the CFA readmitted to FIFA and the way paved for the commercialisation of Chinese football. In response to the disappointing performances of the national team in the 1980s and in view of the creation of 'a new socialist market economy', ever since the early 1990s the NSC was significantly involved in implementing fundamental changes to the football system through encouraging the professionalisation of Chinese football, restructuring the institutional framework and setting up football companies as commercial agents (Jones, 2004: 56–57).

After the foundation of a professional league had been decided in 1992, the CFA was redefined as a non-governmental body while remaining de facto strongly attached to the state, namely the General Administration of Sport (GAS). In 1994 the Chinese Professional Soccer League was put into being, and the first season was launched consisting of two divisions, the Jia-A and the Jia-B, with several professional teams based on a franchising system. In terms of commercialisation, clubs and teams previously owned or operated by governmental agencies were officially attached to enterprises and companies in order to fuel more money into the league. However, in practice most of these enterprises were in fact state-owned and thus 'company sponsorship has been simply a form of public subsidy in another name' (Amara *et al.*, 2005: 194–195). While the league system was partly liberalised in terms of marketing and merchandising, the degree of governmental regulation maintained relatively high, not only with regard to commercial rights but especially in relation to transfers of players by setting up age limits, maximum wages and restrictions on going abroad. However, these have been softened since 2006 and transfers to foreign countries have been encouraged, hoping that the experience players could gain from other leagues would contribute to the quality of the national team (Tan and Bairner, 2010: 591–592).

Despite the fact that by 2001 the league had attracted an overall spectatorship of 3.25 million, and although the national team had qualified for the World Cup 2002, its popularity and appraisal declined rapidly in conjunction with the surfacing of a major match-fixing and corruption scandal, which came to be known as the 'Black Whistle Affair', as well as with dramatic financial instabilities in the club system. Moreover, the CFA failed to foster youth and grassroots football effectively, which impacted on the number of active players and in particular the quality of the national team, to the extent that fans were infuriated and engaged in violence (Zhe Tang, 2011; Jones, 2004: 60–63). People began turning their backs on Chinese football and stadiums were so empty that one could even 'get a ticket for free with buying a pack of cigarettes', probably enhanced by the phenomenon that most spectators had previously been more interested in football as an event, 'pretending to be a fan' rather than being genuinely devoted to the game (Yang Fei, 2010). As a result, all stakeholders engaged in at times conflictual discussions on how to resolve the crisis, finally leading to a restructuring of the league which was renamed the Chinese Super League in 2004.

According to the CFA Development Plan issued at the beginning of the decade, the development of Chinese football has been insufficient considering the overall economic growth and organisational structure of the Chinese sport system. Against this backdrop, the CFA (2003) formulated the following guiding ideology for the years between 2003 and 2012:

> Using the Chinese Super League (CSL) as the lead; youth and coach development as the key point the overall standards of football will be raised and the competitive ability in the international arena will be enhanced. The core objective of CFA is to transform China into one of the strongest footballing countries in Asia. These aims will be achieved by continuous reforms, marketing led strategies, structural adjustments, improved relationships, optimum resource distribution and harnessing the enthusiasm of all quarters of society.

In order to achieve the goal of becoming a first-class footballing nation in Asia, the CFA put focus on the league, the clubs, the infrastructure and the competitions. Moreover, the declaration contains a statement according to which the CFA would strengthen its ties with other east Asian national associations, AFC and FIFA in order to attain the status of 'an active player in the dynamics of global football' (Chinese Football Association, 2003: 10). Indeed, the Asian Cup was hosted in China in 2004 and the Women's World Cup in 2007.

Unfortunately the role of women's football in China turned out to be of marginal significance with regard to increasing the overall popularity and optimising the development of the sport itself, at least in the long run. Despite the early success of the national team in the 1980s, when they won the Asian Women's Football Championships in 1986, thereby paving the way for the foundation of the first professional club in 1990 and the hosting of the first ever Women's World Cup in China in 1991, female players were treated unequally in terms of training facilities and living conditions (Jinxia and Mangan, 2002: 93). Paradoxically, this situation actually motivated some of the players and the Chinese government seemed to be highly appreciative of the sporting results, namely winning the silver medal at the 1996 Olympics and the achievement of reaching the third round at the World Cup 1999, particularly due to the exposure at the global level and the impact it had on Chinese fans who then proudly called the team the 'Iron Roses' (Hong and Mangan, 2003: 47). With the national team's results deteriorating by the beginning of the millennium, and in consideration of the prevalent degree of institutionalised discrimination, 'gender politics, contemporary commercialism and resuscitated traditionalism' (Jinxia and Mangan, 2002: 11ff.), women's football has since experienced a rapid decline both domestically and internationally.

In spite of the ambitions and objectives defined by the CFA, the early twenty-first century continued to be a dark chapter for Chinese football. In fact, another major corruption and match-fixing scandal overshadowed the establishment of the CSL. The government was forced to launch a nationwide investigation in 2006, revealing forms of corruption at almost every level. Amongst others, two former Vice-Presidents were accused of bribery. As a consequence, the TV channel CCTV cancelled its broadcasting of domestic league games. In collaboration with the GAS and the police, the CFA promised to install an effective system aimed at cleaning up corruption, while admitting that it requires a long-time effort given the continuity and scale of the scandals. It is in this context that Chinese football needs to re-establish its image and its attractiveness in commercial terms, as well as to finally find ways of promoting youth development in order to raise the quality of the national team (Soccerex, 2012; Xinhua, 2011). A contract signed with Dalian Wanda in 2011 – covering a sum of US$77.3 million to be invested in youth development, upgrading the domestic league as well as referee and coach education schemes – may serve as a first step (*China Daily*, 2011). In addition, the Chinese Sport Ministry has recognised the importance of the most popular sport on the globe and has generally considered bidding for hosting a World Cup, thereby hoping that the tournament could have an impact on mass participation and renew public interest in local football (Sun Daguang, 2010).

Singapore

The first ever football association to be founded in Asia was the Singapore Amateur Football Association (SAFA) in 1892, which nowadays goes by the name of Football Association of Singapore. British workers who used to work for a company located in Singapore founded it. Given the presence of British companies and their trading activities in the southeast Asian region at that time, it was common that expatriates used to conduct football matches and create teams. British engineers on a pitch near Tank Road played the first recorded match in 1889. In the beginning it was only British playing with and against each other; regimental clubs were founded and civilians came regularly together to play. In the last quarter of the nineteenth century, however, locals increasingly engaged in physical activities and football as a recreational option gained popularity amongst most ethnic groups (Tomlinson, 2005: 131).

In the year of its foundation, SAFA organised its first competition named the Association Challenge Cup. Due to its success and the high demand for regular competitive matches, the governing body decided to inaugurate the Singapore Football League in 1904. Its first winning team was the 1st Battalion, Manchester Regiment. European teams and clubs dominated both competitions but, due to the fact that local teams installed their own communal leagues and organised inter-ethnic friendly matches, they quickly improved their level of play. In 1925 the team of the Singapore Chinese Football Association succeeded in winning the Football League for the first time, followed by the Singapore Malay squad taking the trophy in 1934.

When HMS *Malaya* visited Kuala Lumpur and Singapore in 1921, the HMS Malaya Cup, later to be known as Malaysia Cup, was put into being and took place in Singapore in 1925. Despite its overwhelming popularity, other formats and leagues also emerged, such as the Government Services League and the Business Houses League. While the Singapore Football League experienced a rapid extension of clubs, including expatriate teams, in the 1950s, the Business Houses League became more attractive to corporate clubs such as Singapore Airlines and Malaysian Breweries. In 1952, SAFA was transformed into the Football Association of Singapore (FAS) after cancelling its prior affiliation with the Football Association of Malaya, becoming an independent national association. In 1961 the league system was put on ice until it was relaunched in 1975. Meanwhile, the FAS had been amongst the founding members of AFC in 1954 with Soh Ghee Soon, FAS President from the early 1950s until 1963 and also an AFC Vice-President, being the longest-serving football administrator during the period.

At the international level, Singaporean squads gained only marginal success. Apart from reaching the semi-finals of the Asian Games in 1966, further major victories failed to materialise. At least competitions such as the Merdeka Tournament or the King's Cup in Thailand were frequently contested. Despite the victory over Penang in the Malaysia Cup final in 1977 and football's popularity reaching an unprecedented high, Singaporean football henceforth started to suffer a decline. After the new National Football League (NFL) had been installed in 1975, match-fixing scandals tarnished its reputation considerably. The NFL also struggled, receiving generally less public attention than the Malaysia Cup. However, in spite of the historical importance of the tournament, the FAS later decided do withdraw from the Malaysia Cup, as well as other Malaysian league formats, due to political reasons and fear of corruption. Also aimed at fostering local talents, coaches and teams, the decision was however largely regarded as unpopular. FAS President Mah Bow Tan worked closely with league and club officials on finding a way to deal with the discontent. This resulted in the creation of the S-League in 1995. In addition, football development initiatives were put underway through the establishment of the National Football Academy and the Foreign Talent Scheme. In fact, the Singaporean team was then capable of winning its first international trophy at the Tiger Cup in 1998. Moreover, the FAS and the Singaporean government decided to pass legislation according to which betting was legalised in 1999. This measure was regarded as a means of counteracting the problem of illegal betting, which was closely connected to match fixing and bribery. The additional money generated could also be channelled into developing the league's status (Football Association of Singapore, 2011).

Furthermore, the S-League experienced a modification of its rules in relation to making transfers of foreign players more flexible, and prohibiting players under 18 years old from participating in league matches. Meanwhile, plans exist to revive Singapore's participation in the Malaysia Cup. In collaboration with the Football Association of Malaysia and following a new strategic plan for the time from 2010 to 2015, options for installing an ASEAN Super League have been discussed, a Singaporean team has entered the Malaysian League, and Malaysia's Young Tigers have joined the S-League (Football Asia, 2000; Voon, no year; Chen, 2011).

Malaysia

By analogy with how football found its way to Singapore, the Malayan mainland was also not exempt from colonial forces and thus it was the British who introduced the game in ports like Malacca. By the end of the nineteenth century, football had indeed become one of the central pillars in most sports clubs in Malaya. At that early stage football lacked an elaborate organisational structure, which only took shape in 1905 when the Selangor Amateur Football League was founded, though exclusively consisting of teams from Kuala Lumpur. Notwithstanding, friendly matches were played with teams from other regions and it occurred frequently that teams from Singapore played against Selangor clubs. In 1926 the Selangor Football Association was founded and the league affiliated with it. Ten years later the governing body was renamed to Football Association of Selangor (Football Association Malaysia, 1991: 8ff.).

With the establishment of the Malaya Cup in 1921, football continued spreading throughout the country and soon other regions founded their own associations, such as in Johor and Perak. At the time, the composition of the teams was mainly based on ethnic background, e.g. the Selangor Chinese Recreation Club, while there were also some Eurasian clubs. In the 1920s and 1930s, the local football culture flourished, rivalries emerged and public interest grew. In this context, the Football Association of Malaya (FAM) was founded in 1936, after its predecessor the Malaya Football Association had been founded in 1926. At international level, the Malayan Chinese Olympic team represented the country. Interestingly, the FAM was first located in Singapore before relocating to the mainland after the Second World War. Yet the influence of the Singapore Amateur Football Association remained strong. This changed when SAFA became a national association in its own right in the early 1950s (Seneviratne, 2000: 11–20).

The 1950s marked the beginning of the heyday of Malaysian football. Tunku Abdul Rahman, the first Prime Minister of the newly independent Malaya, was significantly involved in pushing for the construction of the Merdeka Stadium, which was to be the venue for the independence celebration in 1957. It also signalled the launch of the Merdeka Tournament, an international tournament of high prestige for Asian countries from all across the continent (Seneviratne, 2000: 24). The Prime Minister (quoted in Seneviratne, 2000: 30) described the spirit of the first tournament in 1957 as follows:

> The Federation may not come out victorious, but we should at least be happy in the thought that we have arranged the programme not only with the intention of entertaining our football fans but also to promote goodwill, good sporting spirit and the friendship with all the countries in the world.

Indeed, the tournament became a success and Malaya lifted the cup between 1958 and 1960 three times in a row. Off the pitch, meanwhile, Tunku was elected President of the FAM in 1951 and of AFC in 1958. Determined to step up the development of Asian football under his leadership, youth tournaments at the domestic and international levels were installed, coach seminars were conducted, the FAM built a new headquarters in Kuala Lumpur, and the issue of sufficient representation of Asia in FIFA was tackled (Seneviratne, 2000: 34–40).

In the 1960s the creation of Malaysia – now including Sabah, Sarawak and Singapore (the latter left the federation in 1965) – led administrators and politicians to identify new opportunities for reforming the sport system. Accordingly, the FAM decided to no longer promote communal football structures, assuming that as time passed the regional associations would realise that their existence was meaningless (Seneviratne, 2000: 53–56). Meanwhile, the Malaysian government formed a ministry responsible for sport and youth in 1964 and later installed the National Sports Council (NSC), which took over responsibility for elite sports. The NSC consulted with the Olympic Council Malaysia, the national federations and governing bodies in relation to monitoring the processes of sport development. Due to the fact that the NSC was also responsible for funding matters, the federations became reliant on governmental subsidies, leading to the consequence that some of the larger federations recognised the need to retain their financial and political autonomy (Sieh Kok Chi, 2011). The FAM, for instance, became largely self-sustained and governmental involvement was reduced, while NSC representatives would still have seats on the FAM committees (Ahmad, 2011).

Despite the fact that the 1970s saw a very strong Malaysian team – including star players such as Mokhtar Dahari, winning the Merdeka Tournament, and qualifying for the Olympics in 1972 and 1980 – and despite the fact that the overall number of players increased at grassroots level because football had been incorporated into the physical education curricula at schools, public interest and attendance rates declined dramatically (Seneviratne, 2000: 61ff.). In a sense, this signified a break in the positive development of Malaysian football. While the FAM had been long considered the best structured governing body throughout Asia, serving as role model for countries such as Japan, the 1980s turned out to be the period in which in sporting terms Malaysia lost touch with other Asian countries such as South Korea, despite the efforts that had been made through the instalment of a youth development project and 14 centres of excellence in each of the provinces (Samuel, 2008).

Under the presidency of Sultan Ahmad Shah in the mid-1980s, the FAM looked for options to improve the standard of the domestic football scene and decided to launch the Semi-Pro Football League in 1989. The idea behind it was that first only the players were to receive professional status, and later the coaches and administrators. Accordingly, a statute was introduced by which a team had to have at least three contract players. What made the league really unique was the fact that it did not consist of clubs but of 14 states and two countries, namely Brunei and Singapore, with the latter pulling out in 1995. The FAM also founded the Malaysian FA Cup, which however was only designed for Malaysian teams. Unfortunately, the launching of these new formats was inhibited from the very beginning. A dramatic match-fixing and bribery scandal, resulting in a devastating clean-up campaign initiated by the government and the FAM, was detrimental to the advancement of the game. By the mid-1990s, 57 players and one assistant coach were suspended and 32 players and coaches banished, while others escaped prosecution. It has been estimated that bribes totalling RM 1.324 million had been paid (Seneviratne, 2000: 106–111). Back then, FAM Vice-President Tengku Makhota (quoted in Seneviratne, 2000: 116) commented on the incidents as follows:

> The decade of the 1980s was the years the locusts have eaten. Compounding matters was the growing cancer of bribery in Malaysian football. The decision of the FAM to transform the Malaysian League from amateur to semi-professional and then professional status was prompted, in part, to stem the tide of soccer bribery in the late Eighties. However, the move to professionalism though necessary, wasn't effective at banishing the bribery menace.

In fact, the problem of bribery and corruption expanded in spite of the measures taken by governmental authorities. Hence the sport suffered a lot, star players were banned and public interest decreased, while European leagues caught increased attention of the fans, thereby affecting domestic league spectatorship rates (Goldblatt, 2006: 852). While FIFA and AFC put pressure on the FAM with respect to counteracting the match-fixing structures, a new Malaysian Premier League format was introduced in the 1990s, finally resulting in the launch of the Malaysia Super League (MSL) in 2004. Also, the FIFA World Youth Tournament was hosted in Malaysia in 1997, and in this context a new vision was formulated in order to bring Malaysian football back to its better days: the so-called 'Three D's', standing for desire, discipline and determination (Seneviratne, 2000: 114, 127, 158).

In light of the ongoing scandals, however, the FAM has been criticised for delivering no more than cheap talk. The formulated aims of improving administration, development programmes and professionalisation procedures have hardly been realised. Instead, the fragmented administrative structure has been viewed as the reason for the MSL not fulfilling the AFC's requirements for professional leagues and for none of the teams managing to qualify for the AFC Champions League (Samuel, 2011). Also, the FAM has suffered severe financial struggles and funding problems, thus looking for governmental support, while not showing enough collaborative effort to improve football at, for instance, the school level (Ahamd, 2011). On the other hand, the FAM has begun working closely with the NSC through the establishment of a government-funded national development plan ultimately aimed at producing the Under-20 World Champions in 2019, while also looking into how to foster school football and talent development at the regional level (Lim Kim Chon, 2011). Whether such initiatives can one day finally lead to a Malaysian team qualifying for the first time for the World Cup remains to be seen, not least since, according to an FAM official, match fixers severely tarnish the sport in southeast Asia (*The Straits Times*, 2011).

India

In the era of British colonialism, football in India was facilitated by British regimental teams and missionaries. In Victorian public schools in north and east India, football was instrumentalised as a moral tool in order to educate and 'civilise' the local populations (Mangan, 2001: 41; Majumdar and Bandyopadhyay, 2005a: 125). Due to the fact that the British government was located in Calcutta, the first clubs were founded in the eastern region of India during the 1880s. Amongst these were Mohammedan Sporting, founded around 1891, and Mohun Bagan, which was formed in 1889 and has been the only Indian club winning every major domestic tournament, for which it was declared the National Club of India in 1989. Covering various sports such as football, cricket and hockey, Mohun Bagan has ever since been a figurehead of Bengal due to its special ethos, while being supported by the upper class. In the early days players did not receive any payments and most of them were students who the club helped find jobs that gave them enough time to play football (Kapadia, 2001: 17).

Together with the British teams, Mohun Bagan and Mohammedan Sporting began competing in the so-called Calcutta League, while the three major tournaments were the Indian Football Association Shield, which took place in Calcutta, the Rovers Cup in Bombay, and the Durand Cup in Simla. The latter is the oldest Indian competition and in fact the third oldest football tournament in the world, since it was incepted in 1888. Lifting all three trophies in one season was named the Triple Crown and it has been the highest achievement a team could accomplish. A major watershed event in Indian football history was the final of the IFA Shield in 1911 when Mohun Bagan defeated the East Yorkshire Regiment. Its significance derives from the fact that it was the first time an Indian team had beaten a British side and won an important tournament. The Indian nationalist movement perceived the victory as a symbolic victory over the British rulers (Kapadia, 2001: 18; Dimeo, 2001: 107–109).

Moving beyond Calcutta, football spread in other British garrison towns such as Madras, Bangalore, Hyderabad, Delhi and Peshawar. The Bangalore Muslims club, for instance, was the first Indian team to win the Rovers Cup, in 1937, while by the 1940s the international team consisted mainly of players from Mysore. Regardless, the regional powers remained Calcutta and Bengal. Nonetheless, football began to flourish in places like Hyderabad, where the Hyderabad Football Association was formed in 1939. The most famous team from the city was Hyderabad City Police, which won the Durand Cup by beating Mohun Bagan for the first time after India had gained its independence in 1947. By the early 1950s the team had established itself as one of the best in the country (Kapadia, 2001: 19).

The Durand Cup long remained the most prestigious domestic competition and only lost its status in 1996 when the semi-professional National Football League was launched. The tournament had always taken place where the government was located, hence ever since 1947 in Delhi. The first Indian team lifting the trophy was Mohammedan Sporting by winning against the Royal

Warwickshire Regiment in 1940. This victory marked the pinnacle of a tremendously successful decade for the club after it had won the Calcutta League five times in a row in the 1930s, thereby receiving huge popular support from Muslims all across India as well as attracting young players and talents to Calcutta (Kapadia, 2001: 20–22).

Through the foundation of the East Bengal football club in 1924 by fans who had migrated to Calcutta in search of work, Mohun Bagan gained 'a social, political and geographical rival' (Kapadia, 2001: 22). While Mohun Bagan was the older club, supported by the rather wealthy population of west Bengal, East Bengal's support drew upon people with a lower-class and working-class background. The rivalry was indeed so intense that, when Mohun Bagan could not win against East Bengal for a period of six years during the 1970s, a fan committed suicide, hoping this sacrifice would help change the gods' minds about the outcome of future matches (Dimeo, 2001: 114). East Bengal has managed to become an unforgettable part of Indian football through historic victories over the Pas Club of Iran in the 1970 IFA Shield and the triumph over Pyongyang City Club of North Korea in the 1973 final. In addition, East Bengal was to become the team winning more games at home against foreign clubs than any other Indian team (Kapadia, 2001: 22).

Meanwhile, at the international level India made its first major appearance during the 1948 Olympics held in London. Playing barefooted, the team could hardly cope with the bad weather conditions and lost to France. After the Indian team had frozen their feet at the Helsinki Olympics in 1952, the All India Football Federation, which had been founded at a meeting of six state associations in 1937, decided to oblige players to wear boots. Until 1960 India qualified frequently for the Olympics and achieved a memorable fourth place at the Melbourne Olympics in 1956. Meanwhile, the national team also accomplished qualification for the World Cup 1950 in Brazil, but could not participate due to a lack of financial resources and logistical issues, as well as the problem of barefooted players. This affected the further development of Indian football dramatically in spite of winning the Asian Games in 1951 and 1962. The achievements of the heyday period were not repeated and India had to wait until the 1990s, in which the national team won the South Asian Football Federation Championship twice in 1997 and 1999 (Majumdar and Bandyopadhyay, 2005b: 286–289).

Concerning the role of women's football in India, it has to be said that football is still considered mainly a male domain. It is mostly women with a lower-class background who play, and the few top players hardly receive any recognition due to the fact that women's football associations suffer from severe financial lacks. As the average Indian male has never really supported the cause of women's football, because the sport is seen as unsuitable for women who are thought of to rather take care of the household, Indian sports officials have generally neglected the structural and financial problems women's football governing bodies face. Although the origins of women's football in India date back to the time of the emancipation and suffrage movement – in the course of which better sports facilities for women were demanded and the first women's football tournament was held, as well as the Women's Sport Federation established in 1939 – the sport quickly declined, as in Bengal by the early 1940s. Even despite the founding of the Women's Soccer Federation three decades later in 1975 and the establishment of a national league, the situation for females playing the game has almost changed not at all. In 2002, only a few states promoted the growth of the sport and therefore there have been dramatic imbalances, with Manipur dominating the scene ever since the 1990s. Hence it is important to ask whether there will be a viable future of women's football in India (Majumdar, 2003: 81–90).

The structure of Indian football has been characterised by the regional pattern that had been put in place by the British. Initially there had been only one viable governing body, namely the IFA, which was founded in the 1890s in Calcutta and was affiliated with the English FA. In the 1930s, however, other regional associations emerged, such as in Bombay and Delhi, and therefore contested the dominance of Bengal. As a result of the fragmented structure, alongside the political rivalries within east India on the one hand and with the other associations from Delhi and Bombay on the other hand, the All India Football Federation (AIFF) came into being in 1937, which nowadays consists of 36 state associations. It was no easy task for the AIFF to combine the different interests of the state associations but, because the English FA and the Army Sports Control Board of India supported the cause of Delhi and Bombay, Calcutta had to succumb. Regardless, the rivalries between the regions could hardly be tamed and after India had gained its independence they became even bigger, leading on the one hand to a problematic relationship between the club and national levels while also contributing to the overall popularisation of football in India in the 1950s and 1960s, albeit at the cost of administrative improvement as well as the development of India as a footballing nation at the international level. Furthermore, a critical issue was the fact that, after the clubs and associations had initially been exclusively run by British officials, coaches and players, and only later complemented by Indian sports personalities and administrators, leading Indian politicians became involved in running the game. Hence from the late 1940s onwards the AIFF became a playground for personal political rivalries, posts were held without any previous experience in sporting administration and the internal problems led to the AIFF not being capable of providing enough support for the sake of Indian football.

As Indian football experienced its decline in sporting terms in the late 1960s, more politicians became involved in the administration, so that one AIFF President for instance was a full-time politician who did not have the necessary experience but held the post for reasons of social prestige (Bandyopadhyay, 2011). According to the sport historians Majumdar and Bandyopadhyay (2005b: 291), such patterns prevail to the present day and require radical reforms:

> The unresolved dichotomy of national and club football has been, to a great extent, a result of its failure and amateurish duplicity. Moreover, factionalism, favouritism and infighting within the Federation are plain to see since its inception. [...] It is, however, ahistorical to put the entire onus on administrators and the like. In fact, one must acknowledge that it is Priya Ranjan

Dasmunshi, the President, who has almost single handedly kept the National Football League afloat. It was his initiative that roped in financial giants like ONGC [Oil and Natural Gas Corporation] as title sponsor before the National League in December 2004. In fact, it is rather the peculiarly amateur set-up of soccer that explains the deep-seated ills of Indian football. This is not to deny that AIFF officialdom has been the major villain in this system.

The regionalist structure has had far-reaching implications with regard to professionalising the game because most of the time leagues and tournaments merely took place at the regional level. The AIFF had temporarily installed a national championship for the Santosh Trophy from 1941 until 1942, but it took another six decades until the NFL was founded in 1996. Initially consisting of eight teams, the first two seasons were sponsored by Philips and the following two by Coca-Cola. However, the AIFF did not succeed in extending a contract with the TV broadcaster ESPN, nor in ensuring a reasonable degree of financial sustainability due to poor organisation and marketing efforts. While ever since the 1980s corporations and business developed a major interest in investing into football – thereby providing grounds for newly founded fully professional clubs such as FC Kochin and for opening up to the global player transfer system, as well as stimulating a considerable increase in players' salaries – the commercialisation of Indian football underwent a rather swift period of rise and fall. By the 2000/01 season corporate interests had decreased and the three major competitions, the Federation Cup, the Santosh Cup and the NFL, were left without sponsors. As a consequence, the clubs criticised the AIFF for not having managed and dealt with these developments adequately. Complicating the matter was the fact that football found a commercial competitor in cricket, which is still considered the biggest sport in India (Rodrigues, 2001: 106–125). Moreover, the focus on promoting the NFL:

> systematically undermined important and prestigious tournaments such as the Durand Cup, DCM Trophy, Rovers Cup and IFA Shield which used to act as breeding grounds for soccer talent in the country as well as to nourish a sound club culture.
> (Majumdar and Bandyopadhyay, 2005b: 292)

After a new title sponsor had been found in the form of Oil and Natural Gas Corporation Limited in 2004, in order to overcome the crisis the AIFF decided to take Indian football one step further towards full professionalisation through the launch of the I-League in 2007. In spite of a new marketing deal with IMG Reliance, signed in 2010 and covering a period of 15 years, as well as renewed public interest and attendance figures of up to 100,000 during matches between East Bengal and Mohun Bagan, the I-League suffered a severe setback when it was revealed that the league's teams would not participate in the AFC Champions League because the I-League did not fulfil the AFC criteria for 2012 (Cutler, 2010; 2006: 518).

Meanwhile, some of the state associations, such as the Goa Football Association and the IFA, have taken their own measures with regard to fostering the development of the game, which in turn inspired others such as the All Manipur Football Association to initiate youth tournaments, school competitions and other events in order to promote football as an entertainment. The IFA, for instance, launched a state football league and installed a football academy aimed at grassroots and youth development (Majumdar and Bandyopadhyay, 2005b: 294ff.).

Whereas the Bengal clubs have not adapted to the global standards, Goa has come up with a professional approach both in technical and administrative matters, thereby constituting a counterweight to the traditional dominance of Bengal and Panjab as well as Mumbai, from where some clubs have even withdrawn from the national league. The AFC's Vision Asia programme has also helped the northeast regions such as Manipur, raising standards and creating a new generation of footballers and a new football culture. In contrast to these regional attempts, the recent development of the I-League has been viewed very critically. According to Bandyopadhyay (2011):

> It has to be said that it was a great illusion that AIFF officials thought it could follow the footsteps of the EPL or the J-League. It didn't work out because of poor marketing, poor structure and most importantly the AIFF was not run by professional people.[...] Also, the government never realised the potential and importance of sport in modern civilization – that's a fundamental problem. Hence the government never looked at soccer [as] something to be taken care of.

In this sense it will be interesting to see how both the AIFF and the Indian government will engage in promoting football in India, its professionalisation, commercialisation and further development.

Iran

Football found its way to Iran in three ways: through missionary schools, the oil industry and the military. With the foundation of the first modern school in Teheran in 1851, Iranian students were exposed to regular exercises. Physical education as a tool for national progress was publicly discussed by the turn of the century, while competitive western games were brought to Iran by students who had spent time in Europe, as well as by British expatriates. The first recorded football games took place in Ishafan in 1898, where a British team played against an Armenian squad. Whereas football had been part of the curriculum of the missionary schools attended by pupils from the Iranian elite, working-class Iranians got in touch with the game through British workers of the Anglo-Persian Oil Company. In fact, at the site of Masjed Soleyman, a league had been installed and an annual match between England and Scotland organised, while Iranians began replacing British players on the teams before founding their own. In Teheran, meanwhile,

Iranians became involved in football teams around 1908 (Chehabi, 2006: 235–237; Fozooni, 2004: 357–358).

During the First World War organised football had ceased, but it was reactivated by 1918 and two years later the first all-Iranian football club, named Iran Club, was put into existence. Either in 1919 or 1920, Iranian and British members founded the predecessor of today's Football Federation Islamic Republic of Iran (FFIRI) in order to promote the game, naming it *Majma'-e futbal-e Iran*. A year later it was fully taken over by the Iranians who changed the name into *Majma'-e tarvij va taraqqi-ye futbal* (Association for the Promotion and Progress of Football), and it became the first modern sport governing body registered at the State Registry. What followed were publication of the rules in Persian and the organisation of regular tournaments after 1923 (Chehabi, 2006: 237).

As football had emerged as a modern element in Iranian society by the mid-1920s, the significance of the sport was recognised by political leaders. Hence football games became a regular physical activity for the military. Further development of the game as a mass sport was hindered by a severe lack of facilities, while with the accession of Reza Shah in 1925 football continued to be promoted, particularly with regard how it could contribute to the nation's identity. In fact, Reza Shah was delighted to see an Iranian team win for the first time ever against a British selection in 1926. Subsequently an Iranian squad was sent to the Soviet Union in order to help improve the countries' relations and in 1929, when a team from Baku was invited to Teheran, an effort was made to equip the state-owned pitch with grass. Meanwhile, by the beginning of the 1930s, the Association for the Promotion and Progress of Football was replaced by several other bodies. In 1934 the Office of Physical Education was put in place, henceforth being in charge of school football, while later that year the National Association for Physical Education was founded. As a result the first national championships were organised in 1939, including football amongst other disciplines. The following year initial attempts were made to establish specific federations which, however, were not realised until after the Second World War (Chehabi, 2006: 238–240; Fozooni, 2004: 358–360).

When Reza Shad left Iran in 1941, political power remained largely with the modernists so that football was still promoted, albeit less than before. Resulting from the fact that during the Second World War Iranians used to play matches against foreign soldiers, officials recognised the importance of installing appropriate governing bodies, fulfilling the requirements of the international sport scene. Accordingly, a national Olympic committee was put in place in 1947 followed by the establishment of the national football federation, which was affiliated with FIFA in 1948. However, the national team's performances were poor during the 1950s and this changed only by the mid-1960s, when victories over Pakistan, Iraq and India led to the qualification for the Tokyo Olympic Games. Football lagged behind the significance of wrestling as Iran's most popular sport, but in the 1960s it experienced a growth in terms of becoming a spectator sport owing to urbanisation trends that increased the possibility of masses developing an interest in the game. In addition, Iranian television channels began broadcasting games such as the final of the Asian Cup in 1968, in which Iran defeated today's AFC ex-member association Israel. Going down in history as a watershed event, the game was followed by millions of people on TV, while the stadium served as a stage for Iranian spectators with anti-Semitic resentments. In fact, six years later the teams opposed each other once again at the final of the Asian Games and the Iranian regime instrumentalised the tournament as a means of boosting the nation's image (Chehabi, 2006: 240–242; Fozooni, 2004: 360–364).

At a domestic level, rivalries emerged between a club named Taj, which was regarded as being associated with the regime because it was owned by an army general, and Shahin, mainly consisting of members of the intelligentsia. Against a backdrop with political implications, Shahin was dissolved in 1967, but its players founded a new club called Persepolis. The rivalry between the teams continued to dominate the domestic football scene until the outbreak of the revolution. It was enhanced through the establishment of a national league, which kicked off in 1974. Meanwhile, during the last years of the Shah's regime, critics accused the regime of using football to distract from political issues. In fact, one relative of the Shah was the head of the national association (Chehabi, 2006: 242–243).

In the course of the Islamic revolution, the country's sport policies were tarnished and some sports were even abolished, while all competitions for women were suspended because the religious leaders opined that the athletes were not covered sufficiently. All football clubs were nationalised and renamed, hence Taj came to be known as Esteghlal and Persepolis as Piruzi, which however did not affect the intensity of the rivalry. Due to the political leaders' neglect for elite sports and in light of the war with Iraq in the 1980s, the national team's performances suffered a dramatic decline. Nonetheless, the game's popularity amongst young people did not decrease and, while the new regime had cancelled the Takht-e Jamshid Cup, provincial leagues were installed in 1981 whose winners competed in a national championship tournament called Oods Cup.

Interestingly, major football matches served as a means of importing elements of western culture. In the late 1980s Iran's leaders responded to the need to deal with the people's desires and adopted a flexible stance on sports. Accordingly, sports were promoted as health-enhancing activities while coverage on TV took shape by the early 1990s. As a matter of fact, after the war with Iraq had ended the politicians realised the importance of football and in 1989 a new national league was inaugurated, not least due to the quantity of Iranian players who migrated to other countries and the players' discontent about their amateur status. Even one private club was founded, while the other top teams were still affiliated with state-owned companies or other governmental authorities (Chehabi, 2006: 244–247; Fozooni, 2004: 365).

Despite the government's rejection of western consumerism, football in Iran was not exempt from the commercialisation trends taking place around the globe by the early 1990s. Football had become the most popular sport in the country, enhanced through the increased amount of TV coverage, merchandising and marketing activities. In fact, attendances at national team matches or top club rivalries regularly climbed up to more than 100,000. Under these circumstances, the national team's performances improved significantly and the decade turned out to be the most successful period in Iranian football history. Propelled by winning the gold medal at the Asian Games in 1990, the national team played an unprecedented number of games, thereby benefiting not only the FFIRI's income but also the team's quality of play, which culminated in the qualification for the World Cup 1998 in France. Despite of not reaching the knockout stages of the tournament, the team's appearances caught major public interest. Iran's failure to qualify for the World Cup 2002, however, was reacted to with frustrating anger and disappointment, leading to serious riots in the course of which thousands were arrested. The anger was intensified by the fact that for the play-off game against Ireland restrictions had been loosened only for Irish women, who were allowed to attend the match in the stadium (Chehabi, 2006: 248–251; Fozooni, 2004: 366–367; Bromberger, 1998).

While the 1990s were characterised by a new, successful generation representing Iranian football, resulting in the FFIRI's decision in 2000 to introduce a new league format named the Iranian Pro League, the national team's performance began deteriorating by the turn of the millennium (Football Asia, 2002b). The last major success at the international stage was achieved when ranking third at the Asian Cup in 2004 and qualifying for the World Cup 2006 in Germany. Meanwhile, the Futsal team has been one of the best teams in the world. However, several sources of tension have affected the general progress of the domestic football scene. Certain aspects of running the game have been professionalised, but semi-public clubs are in desperate search of financial resources in order to keep up with the general development of the football economy in the Middle East (Karimpour, 2011). The few privatised clubs, such as Persepolis, on the other hand are becoming financially more transparent, albeit at the expense of handing more say to sponsors. And the increased mobility of players has affected the degree of loyalty and identification with the clubs. Moreover, the rivalries between club teams are sites of political and class conflict including various forms of protests 'in an ever-increasing cycle of animosity directed towards the Islamic theocracy.' (Fozooni, 2004: 368). As a matter of fact, the players themselves have made use of their status in order to express their political opinion. In 2009 national team players such as Ali Karimi and Mehdi Mahdavikia protested against the disputed re-election of Mahmoud Ahmadinejad and were then banned by the authorities. This case exemplifies how football in Iran has been a frequent site for either political expression or governmental involvement, which in fact had already been sanctioned by FIFA in 2006 when the FFIRI's affiliation with the world governing body was temporarily put on ice (Tait, 2009; FIFA, 2006).

Another controversial aspect has been the issue of women either attending football matches or playing the game. In Iran, women have been very active in promoting their cause through and in sports. Following the establishment of the

National Women's Sport Association and various other organisations in 1988, women have been increasingly involved in the administrative bodies, thereby fostering the development of women's sports. While this can be regarded as relatively progressive in comparison to other Muslim societies, financial straits, official policies and the male leaders' view on the traditional role of women in society have impeded progress. In fact, in July 1994 almost 500 women were permitted to attend a preliminary match of the Asian Youth Cup taking place in Iran, but the FFIRI withdrew the decision only three days later because some female fans had been accused of approaching the players asking for autographs. However, four years later a group of women was allowed to play a training session on the pitch of the Hejab stadium in Teheran for the first time since the revolution had taken place in the late 1970s. Moreover, indoor football and particularly Futsal has been discovered and successfully played by Iranian women's teams, and since it takes place indoors the players are allowed to wear short-sleeved shirts and shorts.

Eleven-a-side football, meanwhile, suffers from a lack of facilities that the regime considers appropriate. Notwithstanding this, the motivation to develop the sport at the top level is high, and the clubs and the national team receive financial support from the National Olympic Committee and the Government Physical Training Organisation. However, faced with the problematic issue of securing venues where it is possible to play without the hijab, problems with organising international matches abroad has posed a serious detriment to the development of the national team (Fozooni, 2007: 123; Steel and Richter-Devroe, 2003: 315–322; Chehabi, 2006: 248). In 2011, a qualification match for the 2012 Olympics was cancelled by FIFA due to a conflict with the FFIRI regarding appropriate dress (Momeni, 2011). Against this backdrop, west Asian football officials such as Jordan's Prince Ali Bin Al Hussein have been pushing for an amendment of the international rules of the game in relation to permitting the headscarf (Mackay, 2011). Meanwhile, with regard to allowing women onto the stands of the stadiums, AFC has pushed the issue by requiring the FFIRI to provide access for both sexes and to comply with the confederation's Statutes against the backdrop of Iran hosting the AFC Under-16 Championship in 2012. In fact, President Ahmadinejad, a committed football fan, had actually tried lifting the ban in 2006 but did not succeed due to political power struggles within the FFIRI and the resistance of religious conservatives (Dorsey, 2012a).

Saudi Arabia

In spite of being a society in which most forms of western culture have been, to say the least, viewed sceptically, Saudi Arabia has a vivid football culture sustained by its popularity amongst the people and the political as well as royal elite. It may be argued that football, as one of the few western imports, has actually been accepted for political reasons and due to its global dimension. Not only is the development of Saudi Arabian football closely intertwined with the country's oil resources, but also with how sporting success based on the benefits from the oil industry has been used to enhance the political legitimacy of the ruling class (Goldblatt, 2006: 868).

Saudi Arabian football, at least at the top level, has undergone a dramatic transformation ever since the 1970s due to the support of the extended royal family, which has been significantly involved in running the Saudi Arabian Football Association as well as all major clubs. Although the game had entered the country already in the early twentieth century – followed by the foundations of today's top clubs such as Al-Ahli in the 1950s, and the Saudi Arabian Football Federation in 1956 which was affiliated with FIFA in the same year and with AFC in 1959 – there was only one nationwide competition in place, namely the King's Cup. This changed, however, when the decision was taken to install the Saudi League, which kicked off its inaugural season in 1975. It started with eight participating teams, with each club being allowed to register a total of 45 players. Former international star players were recruited, such as Roberto Rivelino who played for Al Hilal from 1978 until 1983. In organisational terms the great financial capacities allowed drawing upon foreign expertise in order to reach international standards. With regard to developing the country's football infrastructure, the General Presidency of Youth Welfare responsible for sports, culture and social activities benefited from the monarchy's support and fascination for the game. In fact, by the mid-1980s Saudi Arabia counted 150 clubs and a considerable number of facilities and stadiums. On the other hand, rumours exist that the financial dependencies have in turn affected outcomes of matches as well as the selection of squads (Sim, 1984; Goldblatt, 2006: 869).

After having won the Asian Cup in 1984 and 1988, it was decided to introduce professionalism to Saudi Arabian football. In 1990 the league format was modified and a two-stage competition introduced. The new system seemed to bear fruit considering the results of the national team throughout the 1990s. Another Asian Cup trophy was brought home in 1994 and the team qualified for four consecutive World Cup tournaments from 1994 until 2006, thereby making Saudi Arabia a regional powerhouse. In the World Cup 1994 it was the first Arabic side to reach the knockout stages and this success was greatly celebrated at home, not least by the country's leaders.

Despite these supposedly constant achievements at the international level and clubs such as Al-Ittihad being amongst the top clubs in Asia, Saudi Arabian football has experienced a decline in recent years and could not sustain the success of the 1990s. This development is mostly based on the fact that, while elite football enjoys an excellent infrastructure, the grassroots level is underdeveloped. The high number of foreign players that were imported has also hindered producing new talents. Adding to these deficits in the sporting arena is the social dimension. While the role of women in sports, either as athletes or as spectators, remains a controversial topic and the poor cannot afford to buy tickets for the matches, the general discontent about the monarchy and political elites has increased amongst the new middle class. In the context of the revolutionary movements taking place in the Arab world and under consideration of the monarchy's involvement, the situation might also affect the future development of

Saudi Arabian football (Goldblatt, 2006: 869–870; Football Asia, 2002a: 194–197; Dorsey, 2012b).

Qatar (World Cup 2022)

The development in Qatar of western sports and football in particular has been a rather recent phenomenon. The first football club named Al-Ahli was founded in 1950 and the Qatar Football Association in 1960. Ever since the nomination of Qatar as host country for the World Cup 2022, the country and its football development have been in the eyes of the international public. At the time of the bid, football was considered to be the most popular sport in Qatar. Other favoured sports, meanwhile, are cricket, golf, tennis, and traditional horse and camel races. In order to popularise football, former top players had been imported to the Qatar Stars League with the intention of laying the first stone for the major objective of the Qatar Football Association: qualification for a World Cup prior to 2022. Yet, in spite of the measures taken, football matches still do not attract as many spectators as desired by football administrators and decision-makers. Some matches are followed by just 300 people. Most fans tend to watch the matches on television, while the average attendance figure in the stadiums counts only 10,000 per match (News.de, 2011; Goldblatt, 2006: 871; Radnege, 2011).

In relation to improving youth and grassroots development, various initiatives have been instigated. Apart from having established facilities such as the Aspire Academy, partnerships with top European clubs have been installed as a basis for the foundation of further football schools and academies in order to force the pace of the development of children and young players. According to Santino Saguto, an Italian football consultant based in Dubai, Qatar has experienced a jolt in recent years. Besides the organisational efforts that had been put into the hosting of the Asian Cup in 2011, which in a sense could be viewed as a preparatory exercise, the decision to award the World Cup to the small country in the Persian Gulf has inspired the key actors to develop new strategies that shall take Qatari football to a higher level. Consequently, the leagues, associations and media entities from the entire region have engaged in creating new concepts and instruments in order to reach the objectives (Dorsey, 2011a).

In economic terms, meanwhile, another objective is the benefit the World Cup could generate through expanding the tourism sector, thereby complementing Qatar's main source of income, namely its oil exports. Ever since the announcement of Qatar hosting the tournament, more foreign companies have begun setting up business initiatives in the region aimed at stimulating the domestic economy and functioning as an incentive for tourism-oriented programmes (Huld, 2010; Böll and Sievert, 2011: 7).

The main reason for the scepticism about the fact that Qatar has been chosen as host country for the World Cup in 2022 are the climatic conditions in which the tournament will take place. In June and July temperatures usually climb up to 45 degrees Celsius. The people responsible for the bidding campaign, however, promise that all imaginable measures will be taken to utilise technological advancements in order to provide players and fans alike with acceptable conditions. Therefore, a small stadium with 500 seats had already been cooled down to 25 degrees Celsius when a FIFA delegation came to inspect the venue. Moreover, the argument according to which World Cups should exclusively take place in countries with 'normal' conditions has been countered by the Qatari organisers by referring to the discriminatory and thus negative impact such logic could have on the image and status of the global game. Another aspect that contributed to the Qatari bid is the concept of producing modularised stadiums that can be easily deconstructed and shipped elsewhere to prevent them from becoming obsolete. In this sense, the Qatari approach can be regarded as pioneering and trust-building with regard to future bids of supposedly chanceless small football nations (Spiller, 2010). Hence the bidding committee has been putting a considerable focus on sustainability: on the one hand, the World Cup and its related climate issue are intended to help raise general awareness for renewable resources in the region and on the other hand it shall have a substantial effect on the development of the sport itself (Nüssli AG, 2010).

In addition to the critique mentioned in relation to the climate, suspicions emerged according to which votes had been bought in order to ensure the election of Qatar as host for the tournament. Amongst others, officials of the Deutsche Fußball Bund, such as former President and current FIFA Executive Committee member Theo Zwanziger, requested a review of the case (Welt Online, 2011). In social and cultural terms, meanwhile, critics have drawn attention to the fact that Qatar as a Muslim society with its specific norms and values may constitute a setting leading to a culture shock for both the locals and the visitors. This has been exemplified particularly with regard to the problem that alcohol is prohibited in Qatari stadiums, which is hardly understandable for the average European football fan (Focus, 2011).

Other voices from western countries have also expressed their discontent about the way Qatar deals with human rights issues, labour conditions as well as freedom of speech. In the United States there has been a petition demanding the US government takes the World Cup in Qatar as an opportunity to refer to the problem of human trafficking in the region, thereby hoping that global public opinion may put pressure on the actors involved in the criminal actions (Dorsey, 2011b). In relation to the labour policies of Qatar, some activists have already pointed out in the course of the bidding period that the country should try to increase employment rates for the local community instead of outsourcing the labour to foreign workers. This has led to more and more locals having reservations on the project, in particular deriving from the fear that the significant degree of migration may have negative impact on the cultural and traditional structures of their society. The Qatari key actors have reacted to these circumstances by investing not only into World Cup-related infrastructure but also into the installation of educational institutions such as universities in order to foster the education of young Qatari. As a matter of fact, Hassan Al-Thawadi, the General Secretary of the Local Organising Committee, has publicly announced that the organisation of an event such as the World Cup could be confronted not only by criticism and logistical or cultural challenges, but may also even be endangered by natural disasters. On the other hand, he concluded, it could present a platform for building bridges between different cultures and traditions, thereby bringing the west and east closer together (Warshaw, 2011). In this sense, James M. Dorsey (2011a), the author of a blog on football and politics in the Middle East, mentioned that while:

> Qatar and the region on a broader scale are interested in primarily putting the countries on the map and creating an environment to punch upon the weight and thereby influence the public opinion as with shifting it away from orientalist images of the region [...] Qatar 2022 is another incentive in enforcing professionalization and it will sooner or later enhance Asian leagues and therefore the sport-political power of Asian countries in the global governance.

Meanwhile, the Indian sports historian Kausik Bandyopadhyay (2011) has expressed the view that 'Qatar as World Cup host has been a very important choice with regard to boosting and developing the game. In the Middle East football has always been big but this gives something back and may proliferate social energies.' Whether these assumptions will be verified remains to be seen, but the fact that Qatar has been chosen as host for the World Cup underlines the trend that major sports events are increasingly allocated to countries most people would not necessarily associate with a genuine tradition of western sports, thereby indicating that power relations and commercial interests have indeed entered a new era.

Interim conclusions

The challenges that face Asian football and the various stakeholders involved can be better understood under consideration of the historical context and the different political, cultural and economic scenarios in which football has emerged and continues to develop. It is difficult to make general statements when talking about Asian football, and in fact it requires differentiating between sub-regions, countries and their respective conditions, while at the same time certain region-wide phenomena and patterns allow macro-analytical conclusions. Therefore, it is all the more interesting to look at how the responsible confederation deals with handling the diversity and specificities of the particular member associations as well as how these define the scope of duties of AFC on the one hand, and how by the very nature of its mandate AFC constitutes a forum and unified voice for Asian football on the other.

In 1984, then AFC General Secretary Peter Velappan identified four major problems with regard to Asian football, namely the dramatic discrepancy between the countries in terms of performance quality, the vast distances separating them from each other, the lack of professionalism, and the status of the players. Ten years later he stated that these problems have been insufficiently

70 *Developments and challenges*

minimised and that the overall progress in the last 40 years had been rather modest. He concluded that it was therefore utterly important to exploit the commercial potential of the game and to strive towards professionalism in order to reach international standards (Sugden and Tomlinson, 1998: 177). Regarding the diversity of the continent and the confederation, which includes contrasting countries such as the Northern Mariana Islands and India, Velappan (quoted in Sugden and Tomlinson, 1998: 157) mentioned that:

> Asia is a fascinating continent. There is a tremendous diversity of culture, tradition, religion as well as political and social systems. This diversity in a continent that also holds 50% of the world's population gives it a unique strength. In recent years Asia has forged ahead economically. Asia accounts for 25% of global imports and has 33% of global international reserves. By the turn of the century there is expected to emerge a new Asia – and for football unlimited possibilities.

He added that such progress would of course require professionally trained personnel as well as different approaches depending on the particular sub-region or country in question. Accordingly, strategies in the petrol-rich oil states in west Asia would have to be different from those in southeast Asia, where resources are rather scarce and administration has been at times inefficient and corrupt, while the mighty states in east Asia had distinguished themselves through leadership skills, the necessary resources, and popular support for the game. Furthermore, the degree of diversity had increased with the admission of the five former Soviet republics (Kyrgyzstan, Tajikistan, Turkmenistan, Uzbekistan and Kazakhstan) in 1994 (Sugden and Tomlinson, 1998: 157).[1]

In 1995, the Australian sports historian Bill Murray categorised the major challenges in Asian football when pointing to the problems of corruption and violence, the low standard of administration, the weak performances of players, the ambiguous role of broadcasting world football, the huge distances between the countries, the problem of match fixing and gambling, as well as the failure to draw upon the popularity of the game and transform it into an international success (Murray, 1995: 140ff.). As for how critical the problem of match fixing and corruption is and the importance to work against it, the journalist Declan Hill (2011) has referred to the size and impact of criminal networks operating across Asia and other regions, while stating that the lack of professionalisation in conjunction with the criminalisation of gambling has created a gateway for such activities. He therefore suggested that governing bodies such as AFC should install integrity units and foster links with police forces, while ensuring professionalisation and proper payment of players, coaches and referees, as well as providing more preventative education schemes.

Bearing these perspectives in mind, it is therefore interesting to ask what has indeed changed and what has not, or rather how the potential of Asia's diversity and economic progress, as alluded to by Peter Velappan, has been utilised in relation to developing football. It would be naïve to suggest that nothing has changed. The commercialisation of Asian football has indeed been pushed forward. AFC has signed several marketing deals with its partner World Sport Group, formerly known as AFC Marketing Limited, from the early 1990s until today. The AFC Champions League has been optimised following the example of its European equivalent. Countries such as Japan and South Korea in the east as well some of the Middle Eastern states have embraced the commercial value of football quite impressively. Also, more games are broadcasted on TV. Moreover, most of AFC's member associations have committed themselves to professionalising the domestic leagues. As a result, the performances of some of the national teams have improved, not least South Korea's outstanding results at the World Cup 2002 which left a mark. However, with regard to all the abovementioned advancements, one has to consider that the discrepancies between the countries remain strong, that the quality and the sustainability of the measures taken should not be taken for granted, and that further progress requires more time.

When asked, Peter Velappan (2011) referred to ongoing deficits, particularly with regard to the quality of leadership amongst Asian football officials and the administrative bodies. He pointed out that furthermore it is essential to expand initiatives in the technical and educational sector and to tackle the prevalent issues of corruption and match fixing through good partnerships with governments and watchdog agencies, while good education schemes for players and officials could help keep them from getting involved in criminal actions. Meanwhile, AFC Deputy General Secretary Hasan Al Sabah (2011) mentioned that the challenges to Asian football are defined by, first, the size of the continent, which complicates matters regarding the design of match calendars and travel through different time zones; second, the different levels of development; third, the scale of professionalisation, considering that some associations and clubs do not have people working full-time; fourth, the area of marketing, which still requires full exploitation; fifth, how to deal with new technologies in relation to coaching and talent identification; and finally, match fixing, which should be countered especially by means of educating players, coaches and referees in collaboration with other confederations and FIFA. He generally considered the most important element in the whole development process to be the role of the coach, and this in turn requires educational measures, sufficient funding and appreciation. In addition, he stressed the importance of creating links with schools and universities for spreading 'know-how' and providing educational courses for administrators, managers and marketing personnel.

The former AFC Grassroots and Youth Director, Thomas Flath (2012), complemented these diagnoses when saying that generalising on the current situation of Asian football is virtually impossible due to the fact that the different subregions are huge. Countries such as Qatar and the United Arab Emirates, where a lot is happening at the moment, differ from China and India by demographic aspects. Also, the complexity derives from the fact that one would find relatively highly developed football countries like Japan or South Korea in contrast to a lot of actual developing countries, which struggle in political, economic and cultural terms. With regard to enhancing professionalisation, Flath drew attention to the problem that although most countries would actually have a league, often the small number of matches would not be sufficient enough to reach a considerable level of play. Moreover, the fact that clubs were not clubs in a European sense but rather owned by companies or other organisations would explain the severe lack of identification with the teams. Asked about the need to harmonise certain criteria for professionalisation purposes, he referred to the difficulty that in Asia one would find smaller associations asking for hands-on assistance and implementation by a strong confederation, despite envisaging ownership and sustainability, while the bigger associations are interested in maintaining their autonomy. It is this, according to Flath, that clearly separates the situation of AFC from others such as UEFA.

FIFA Development Manager David Borja (2012), meanwhile, emphasised the need to not focus too much on Europe as a role model, and that Asian football should rather engage in an organic process of building its own identity and culture in order to develop the game and increase the popularity of domestic leagues. Elaborating on this aspect, sports journalist Nazvi Careem (2011) expressed the opinion that:

> the countries that manage to develop an own distinct football culture will succeed, because you need fan support. Of course there is the money in Europe, which cannot be easily matched, but at least an own culture would be more than helpful, so when Liverpool come over they want Gamba Osaka to win and not the other way around. It is a fine theory that has been going on for 20 years that European football on TV helps creating a general interest but it has in fact decreased the interest in the local teams at the same time.

Another journalist and former editor of AFC's official magazine, Michael Church (2011), put the spotlight on AFC with regard to the media and marketing dominance of European football when stating:

> The major problem is that AFC itself in fact doesn't play a role in the media landscape in terms of having a grand strategy and actively promoting Asian football. Nobody is in charge of going out and making a bark. There is no promotional drive to say 'look this is Asian football'! A game in China is certainly not as good as a game between Manchester and Liverpool, nobody would have to say that it is as good as in Europe, but nonetheless. The problem is that Telecom Malaysia invests into Manchester United and that money goes out of the local sporting economy. That exploitation is counterproductive. Of course AFC cannot prevent companies from doing that but supporting it doesn't help either and it devalues the commercial opportunities for Asian countries. [...] When Tony Fernandes invests into QPR rather than Selangor, it simply doesn't help Asia. The money that is going out is missing to develop the local game. The assumption that broadcasting international leagues helps the Asian game is a fallacy, especially in countries

like Malaysia where there is a big history. The problem there was of course match fixing and people believe rather in the integrity of EPL.

This statement implicitly brings up many of the above-mentioned aspects that constitute the complex problem that Asian football, whether at the club or the international level, continues to be not an immediate competitor to the big footballing countries and regions, despite the ambitious visions formulated by Asian football officials. For all is intertwined with each other and manifests itself in different variations according to the particular case, situation, as well as location: the continent's size and diversity; the level of play; the tension between popular interest in the national teams versus the local club teams; the league structures and degree of professionalisation; the objective to further commercialisation and marketing efforts against the backdrop of the popularity of European leagues; the development of grassroots and youth football; the education of coaches, referees and administrators; the tension between regulation and autonomy; the interests of member associations and their individual representatives; the roles of governments and other actors; the status of women's football and its promotion; the issue of sport and religion; the pervasive danger stemming from corruption and match fixing; and finally the significance of the continent's confederation in facing these challenges and meeting the demands of Asian football in a globalised game, and whether its ambitious slogan 'The Future is Asia' can be deemed to be an accurate perspective under consideration of the political context in which AFC's actions unfold.

Note

1 Kazhakstan resigned from AFC in 2001 in order to join UEFA. Meanwhile, the Football Federation Australia became a member of the AFC in 2005, thereby adding to the diversity of the confederation's composition.

References

Ahmad, Mazlan (Director General of the Malaysian National Sports Council 1993–2005) interviewed by B. Weinberg, 8 October 2011, Kuala Lumpur.
Al Sabah, Hasan (AFC Deputy General Secretary and Director of Education) interviewed by B. Weinberg, 7 October 2011, Kuala Lumpur.
Amara, M., Henry, I., Liang, J. and Uchiumi, K. (2005). 'The governance of professional soccer: Five case studies – Algeria, China, England, France and Japan.' *European Journal of Sport Science*, 5(4), 189–206.
Asian Football Confederation. (1986). *AFC News*, 7 (No 3).
Asian Football Confederation. (2004). *The power of dreams: 50 years of the Asian Football Confederation*. Kuala Lumpur.
Asian Football Confederation. (2010). *AFC Extraordinary Congress 2010: Agenda 2010 and minutes of AFC 23rd Congress 2009*.
Bandyopadhyay, Kausik (sports historian at the Department of History, West Bengal State University) interviewed by B. Weinberg, 22 September 2011, telephone interview.

Borja, David (FIFA Development Manager) interviewed by B. Weinberg, 1 February 2012, Zurich.
Böll, M. and Sievert, N. (2011). 'Katar im Fokus: Chancen und Projekte zur WM 2022.' Retrieved from www.ostwestfalen.ihk.de/uploads/media/Katar-im-Fokus.pdf on 6 May 2012.
Bromberger, C. (1998). 'Sport as a touchstone for social change: A third half for Iranian football.' *Le Monde diplomatique* (online). Retrieved from http://mondediplo.com/1998/04/04iran on 9 June 2011.
Butler, O. (2002). 'Getting the games: Japan, South Korea and the co-hosted World Cup.' In J. Horne and W. Manzenreiter (eds), *Japan, Korea and the 2002 World Cup* (pp. 43–55). London: Routledge.
Careem, Nazvi (sports journalist and AFC Media Director 2008–2011), interviewed by B. Weinberg, 1 October 2011, Kuala Lumpur.
Chehabi, H. E. (2006). 'The politics of football in Iran.' *Soccer & Society*, 7(2), 233–261.
Chen, F. (2011). 'S-League chief Lee to step down from post.' Retrieved from www.straitstimes.com/BreakingNews/Sport/Story/STIStory_692589.html on 18 September 2011.
China Daily. (2011). '500m yuan to boost Chinese soccer.' Retrieved from www.chinadaily.com.cn/sports/2011-07/04/content_12826541.htm on 16 July 2011.
Chinese Football Association. (2003). *CFA Development Plan (2003–2012) summary*.
Chung Hongik. (2004). 'Government involvement in football in Korea.' In W. Manzenreiter and J. Horne (eds), *Football goes east: Business, culture and the people's game in China, Japan and South Korea* (pp. 117–130). London: Routledge.
Church, Michael (sport journalist and Editor of *AFC Magazine* 1995–2001), interviewed by B. Weinberg, 26 September 2011, telephone interview.
Cutler, M. (2010). 'All India Football signs $155m sponsorship deal with IMG-Reliance.' Retrieved from www.sportbusiness.com/news/182716/all-india-football-signs-155m-sponsorship-deal-with-img-reliance?utm_source=feedburner&utm_medium=feed&utm_campaign=Feed%3A+sportbusiness+%28Sportbusiness+Newsfeed%29 on 19 February 2011.
Dimeo, P. (2001). '"Team loyalty splits the city into two": Football, ethncity and rivalry in Calcutta.' In G. Armstrong and R. Giulianotti (eds), *Fear and Loathing in World Football* (pp. 105–118). Oxford: Berg.
Dorsey, J. M. (2011a). 'Marketing the goal for Middle East football.' Retrieved from www.thenational.ae/thenationalconversation/industry-insights/media/marketing-the-goal-for-middle-east-football on 6 May 2012.
Dorsey, J. M. (2011b). 'Online petition asks Qatar to fight human trafficking in advance of World Cup.' Retrieved from http://bleacherreport.com/articles/608775-online-petition-calls-on-qatar-to-fight-human-trafficking-in-advance-of-world-c on 6 May 2012.
Dorsey, J. M. (2012a). 'AFC puts Iran on the spot on women's rights.' Retrieved from http://mideastsoccer.blogspot.de/2012/05/afc-puts-iran-on-spot-on-womens-rights.html on 22 May 2012.
Dorsey, J. M. (2012b). 'Saudi Arabia.' Retrieved from http://mideastsoccer.blogspot.de/search/label/Saudi%20Arabia on 25 May 2012.
Dunkel, C. (2009). *Ein Mega Event für Japan und Korea: Die gemeinsame Fußballweltmeisterschaft des Jahres 2002 als Spiegel des Verhältnisses der beiden Länder*. Dissertation, Humboldt University Berlin, Faculty of Philosophy, published on 28 May 2010. Retieved from http://edoc.hu-berlin.de/dissertationen/dunkel-carolin-2009-02-13/PDF/dunkel.pdf on 18 February 2015.

East Asian Football Federation. (2005). *EAFF Official Guidebook.*
East Asian Football Federation. (2011). 'The history of East Asian football.' Retrieved from www.eaff.com/fanzone/history/index.html on 17 September 2011.
Eisenberg, C. (2004). 'Fußball als globales Phänomen: Historische Perspektiven.' *Aus Politik und Zeitgeschichte: Beilage zur Wochenzeitung 'Das Parlament'*, (26), 7–15.
Eisenberg, C. (2007). ‚Der Weltfußballverband FIFA und das 20. Jahrhundert: Metamorphosen eines Prinzipienreiters.' In J. Mittag and J.-U. Nieland (eds), *Das Spiel mit dem Fußball: Interessen, Projektionen und Vereinnahmungen* (pp. 219–236). Essen: Klartext Verlag.
Eisenberg, C., Lanfranchi, P., Mason, T., and Wahl, A. (2004). *100 years of football: The FIFA centennial book.* London: Weidenfeld & Nicolson.
FIFA. (no year). 'The history of football.' Retrieved from www.fifa.com/classicfootball/history/game/historygame1.html on 18 April 2012.
FIFA. (2006). 'FIFA suspendiert den iranischen Fussballverband.' Retrieved from http://de.fifa.com/worldfootball/releases/newsid=107802.html on 16 February 2012.
Flath, Thomas (AFC Director of Grassroots and Youth 2003–2006) interviewed by B. Weinberg, 13 February 2012, Mönchengladbach.
Focus. (2011). 'WM-Vorgeschmack in Katar: Kein Bier, keine Fans.' Retrieved from www.focus.de/sport/fussball/wm-2014/wm-wm-vorgeschmack-in-katar-kein-bier-keine-fans_aid_589152.html on 6 May 2012.
Football Asia. (2000). 'New rules for S-League.' *Football Asia: The official magazine of the Asian Football Confederation*, p. 7.
Football Asia. (2002a). *Asian Football Confederation media guide 2002.* Hong Kong: Worldsport Publishing.
Football Asia. (2002b). 'Grounds for improvement: The Iran Pro League.' *Football Asia. AFC Newsletter*, pp. 44–45.
Football Association Malaysia. (1991). *Football in Malaysia.* Kuala Lumpur: Football Association Malaysia.
Football Association of Singapore. (2011). 'History of football.' Retrieved from www.fas.org.sg/fas/history-football on 16 September 2011.
Fozooni, B. (2004). 'Religion, politics and class: Conflict and contestation in the development of football in Iran.' *Soccer & Society*, 5(3), 356–370.
Fozooni, B. (2007). 'Iranian women and football.' *Cultural Studies*, 22(1), 114–133.
Giulianotti, R. and Robertson, R. (2007). 'Recovering the social: Globalization, football and transnationalism. *Global Networks*, 2(7), pp. 166–186.
Giulianotti, R. and Robertson, R. (2009). *Globalization and football* (1st edition). Los Angeles: SAGE.
Goldblatt, D. (2006). *The ball is round: A global history of football.* London: Viking.
Guttmann, A. (1994). *Games and empires: Modern sports and cultural imperialism*, New York: Columbia University Press.
Hill, Declan (journalist and author of the book *The fix: Soccer and organized crime*), interviewed by B. Weinberg, 15 November 2011, telephone interview.
Höft, M., Cremer, M., Embach, C., Jürgens, D. and Thaler, C. (2005). 'Markteintritt europäischer Fußballvereine in Asien: Chancen, Risiken und Handlungsempfehlungen.' In M. Wehrheim (ed.), *KulturKommerz: Vol. 12: Marketing der Fußballunternehmen: Sportmanagement und professionelle Vermarktung hrsg. von Michael Wehrheim* (pp. 145–198). Berlin: Erich Schmidt.
Hong, F. (2002). 'Epilogue: Into the future: Asian sport and globalization.' *International Journal of the History of Sport*, 19(2, 3), 401–407.

Hong, F. and Mangan, J. A. (2003). 'Will the "Iron Roses" bloom forever? Women's football in China: Changes and challenges.' *Soccer & Society*, *4*(2–3), 47–66.

Hoon Han (KFA Coach Education Manager) interviewed by B. Weinberg, October 2011, email interview.

Horne, J. (1999). 'Soccer in Japan: Is *wa* all you need?' *Culture, Sport, Society*, *2*(3), 212–229.

Horne, J. and Manzenreiter, W. (2007). 'Gefangen zwischen Kommerz und nationaler Politik? Der Aufstieg des Fußballs in Ostasien als Re-sultat globaler, nationaler und lokaler Prozesse.' In J. Mittag and J.-U. Nieland (eds), *Das Spiel mit dem Fußball: Interessen, Projektionen und Vereinnahmungen* (pp. 237–280). Essen: Klartext Verlag.

Huld, S. (2010). 'Gastgeber der Fußball-WM 2022: Katar – reicher Zwerg in der Wüste.' Retrieved from www.stern.de/politik/ausland/gastgeber-der-fussball-wm-2022-katar-reicher-zwerg-in-der-wueste-1630583.html on 6 May 2012.

Ichiro, H. (2004). 'The making of a professional football league: The design of the J-League system.' In W. Manzenreiter and J. Horne (eds), *Football goes east: Business, culture and the people's game in China, Japan and South Korea* (pp. 38–53). London: Routledge.

Japan Football Association. (2005). 'The JFA declaration 2005.' Retrieved from www.jfa.or.jp/eng/declaration2005/index.html on 1 December 2011.

Jinxia, D. and Mangan, J. A. (2002). 'Ascending then descending? Women's soccer in modern China.' *Soccer & Society*, *3*(2), 1–18.

J-League. (2011). 'The J-League story: Our place in the football family.' Retrieved from www.j-league.or.jp/eng/organisation/index_02.html on 1 December 2011.

J-League employee (anonymous) interviewed by B. Weinberg, 4 October 2011, email interview.

Jones, R. (2004). 'Football in the People's Republic of China.' In W. Manzenreiter and J. Horne (eds), *Football goes east: Business, culture and the people's game in China, Japan and South Korea* (pp. 54–66). London: Routledge.

Kapadia, N. (2001). 'Triumphs and disasters: The story of Indian football, 1889–2000.' *Soccer & Society*, *2*(2), 17–40.

Karimpour, Amirhossein (employee, FFIRI International Relations Department and AFC Vision Asia Officer 2009–2011) interviewed by B. Weinberg, October 2011, email interview.

Koh, E. (2003). 'Chains, challenges and changes: The making of women's football in Korea.' *Soccer & Society*, *4*(2–3), 67–79.

Korea Football Association. (2011). 'History.' Retrieved from www.kfa.or.kr/eng_renew/library/history.asp on 16 September 2011.

Lim Kim Chon (AFC Technical Director 1999–2004) interviewed by B. Weinberg, 3 October 2011, Kuala Lumpur.

Lee, J. Y. (2002). 'The Development of Football in Korea.' In J. Horne and W. Manzenreiter (eds), *Japan, Korea and the 2002 World Cup* (pp. 73–88). London: Routledge.

Mackay, D. (2011). 'New FIFA Vice-President promises to find solution to Iran dress row.' Retrieved from www.insideworldfootball.biz/worldtournaments/olympics/9329-new-fifa-vice-president-promises-to-find-solution-to-iran-dress-row on 7 June 2011.

Majumdar, B. (2003). 'Forwards and backwards: Women's soccer in twentieth-century India.' *Soccer & Society*, *4*(2–3), 80–94.

Majumdar, B. and Bandyopadhyay, K. (2005a). 'From recreation to competition: Early history of Indian football.' *Soccer & Society*, *6*(2), 124–141.

Majumdar, B. and Bandyopadhyay, K. (2005b) 'Looking beyond the sleeping giant syndrome: Indian football at crossroads', *Soccer & Society*, 6(2), 285–302.
Mangan, J. A. (2001). 'Soccer as moral training: Missionary intentions and imperial legacies.' In P. Dimeo and J. Mills (eds), *Sport in the global society: Soccer in south Asia: Empire, nation, diaspora* (pp. 41–56). London: Frank Cass Publishers.
Manzenreiter, W. (2003). 'Wenn der Zirkus die Stadt verlassen hat: ein Nachspiel zur politischen Ökonomie der Fußball-WM 2002 in Japan.' *Japan 2003: Politik und Wirtschaft*, 223–243.
Manzenreiter, W. (2004a). 'Japanese football and world sports: Raising the global game in a local setting.' *Japan Forum*, 16(2), 289–313.
Manzenreiter, W. (2004b). 'Sportevents und makroökonomische Effek-te: Theorie und Praxis am Beispiel der Fußball WM 2002.' *Kurswechsel*, (2), 67–78.
Manzenreiter, W. (2008). 'Football in the reconstruction of the gender order in Japan.' *Soccer & Society*, 9(2), 244–258.
Manzenreiter, W. and Horne, J. (2007). 'Playing the post-Fordist game in/to the Far East: The footballisation of China, Japan and South Korea.' *Soccer & Society*, 8(4), 561–577.
Momeni, N. (2011). 'Banned Iran women's national team are unfortunate hostages of a political hijab issue.' Retrieved from www.goal.com/en/news/1717/editorial/2011/06/09/2522891/banned-iran-womens-national-team-are-unfortunate-hostages-of-the- on 9 June 2011.
Murray, B. (1995). 'Cultural revolution? Football in the societies of Asia and the Pacific.' In S. Wagg (ed.), *Sport, politics, and culture: Giving the game away. Football, politics, and culture on five continents* (pp. 138–162). London, New York: Leicester University Press.
News.de. (2011). 'Neue Sportmacht: Sport-Wüste Qatar.' Retrieved from www.news.de/sport/855181612/sport-wueste-qatar/1/ on 6 May 2012.
Nüssli AG. (2010). 'FIFA Weltmeisterschaft 2022: Bewerbung Katar.' Retrieved from www.nussli.com/projekte/projekt-detail/news/fifa-weltmeisterschaft-2022-bewerbung-katar-1417/%20334.html on 30 April 2012.
People's Daily. (2004). 'FIFA boss hails China as football birthplace.' Retrieved from http://en.people.cn/200407/16/eng20040716_149849.html on 18 February 2015.
Radnege, C. (2011). 'Is the Middle East a victim of soccer snobbery?' Retrieved from http://christianrad.wordpress.com/2011/08/03/is-the-middle-east-a-victim-of-soccer-snobbery/ on 6 May 2012.
Ravenel, L. and Durand, C. (2004). 'Strategies for locating professional sports leagues: A comparison between France and Korea.' In W. Manzenreiter and J. Horne (eds), *Football goes east: Business, culture and the people's game in China, Japan and South Korea* (pp. 21–37). London: Routledge.
Ritter, M. (no year). *Der Wettbewerb um die Fußballweltmeisterschaft 2002*. Ruhr-Universität Bochum.
Rodrigues, M. (2001) 'The corporates and the game: Football in India and the conflicts of the 1990s.' *Soccer & Society*, 2(2), 105–127.
Samuel, P. M. (AFC General Secretary 2007–2009) interviewed by B. Weinberg, 5 August 2008, Kuala Lumpur.
Samuel, E. (2011). 'Change your affiliates' attitude first, FAM.' Retrieved from http://football.thestar.com.my/story.asp?file=/2011/5/1/football_latest/8588060&sec=football_latest on 18 September 2011.
Seneviratne, P. (2000). *History of Football in Malaysia 1900–2000*. Petaling Jaya: PNS Publishing.

Sieh Kok Chi (Hon. Secretary, Olympic Council Malaysia) interviewed by B. Weinberg, 4 October 2011, Kuala Lumpur.

Sim, J. (1984). *Asian Football Companion.* Unknown publisher.

Soccerex. (2011). 'Korean FA fearful over spectre of corruption.' Retrieved from www.soccerex.com/industry-news/korean-fa-fearful-over-spectre-of-corruption/ on 12 December 2011.

Soccerex. (2012). 'China's anti-corruption trials swing into action.' Retrieved from www.soccerex.com/industry-news/chinas-anti-corruption-trials-swing-into-action/ on 16 February 2012.

Somerford, B. and Kim, Y. (2011). 'K-League set for radical changes with promotion–relegation to be introduced as well as a "split system".' Retrieved from www.goal.com/en/news/14/asia/2011/10/05/2697324/k-league-set-for-radical-changes-with-promotion-relegation-to-be- on 11 December 2011.

Spiller, C. (2010). 'Fußball-WM 2022: "Katar ist ein Vorreiter".' Retrieved from www.zeit.de/sport/2010-12/katar-wm2022-vision-muenchen2018/komplettansicht?print=true on 30 April 2012.

Steel, J. and Richter-Devroe, S. (2003). 'The development of women's football in Iran: A perspective on the future for women's sport in the Islamic Republic.' *Iran: Journal of the British Institute of Persian Studies, 41*, 315–322.

Sugden, J. and Tomlinson, A. (1998). *FIFA and the contest for world football: Who rules the peoples' game?* Cambridge: Polity Press.

Sun Daguang (Chinese Sport Ministry/Director General of the China Sports Culture Development Center) interviewed by B. Weinberg, 6 December 2010, Chengdu.

Tait, R. (2009). 'Iran bans election protest footballers.' Retrieved from www.guardian.co.uk/world/2009/jun/23/iran-football-protest-ban on 20 May 2012.

Tan, T.-C. and Bairner, A. (2010). 'Globalization and Chinese sport policy: The case of elite football in the People's Republic of China.' *The China Quarterly, 203*, 581–600.

The Straits Times. (2011). 'Malaysia FA says fixers "destroying" sport.' Retrieved from www.straitstimes.com/BreakingNews/Sport/Story/STIStory_680498.html on 16 July 2011.

Tomlinson, A. (2005). *Sport and leisure cultures.* Minneapolis: University of Minnesota Press.

Velappan, Peter (AFC General Secretary 1978–2007) interviewed by B. Weinberg, 11 October 2011, Kuala Lumpur.

Voon, T. (no year). 'Malaysia Cup return: Timing "was right".' Retrieved from www.straitstimes.com/BreakingNews/Sport/Story/STIStory_690502.html on 16 July 2011.

Warshaw, A. (2011). 'Qatar 2022 World Cup will bridge gap between east and west, claims top official.' Retrieved from www.insideworldfootball.biz/worldcup/qatar/9922-qatar-2022-world-cup-will-bridge-gap-between-east-and-west-claims-top-official on 6 May 2012.

Welt Online. (2011). 'Korruption bei der Fifa: Zwanziger für Überprüfung der WM-Vergabe an Katar.' Retrieved from www.welt.de/sport/fussball/article13406943/Zwanziger-fuer-Ueberpruefung-der-WM-Vergabe-an-Katar.html on 2 May 2012.

Xinhua. (2011). 'CFA pledges to fight against corruption.' Retrieved from www.chinadaily.com.cn/sports/2011-12/19/content_14288914.htm on 20 December 2011.

Yang Fei (lecturer at Chengdu Sports University) interviewed by B. Weinberg, 9 December 2010, Chengdu.

Zhe Tang. (2011). 'The lost decade.' Retrieved from www.chinadaily.com.cn/sports/2011-10/11/content_13868014.htm on 11 December 2011.

4 AFC's history and origins

The historical development of AFC as an organisation can be viewed in a wider sense as a reflection of and indicator for the development of football in Asia, because it underwent a significant transformation from being a small office in the 1950s to the professionally run governing body that it is today.

The first steps towards installing a confederation for the Asian football associations were taken in 1952, when representatives of the participating Asian countries convened in order to discuss the necessity of creating an administrative body. At the time football leagues and national systems not only in Asia were far from being as elaborately run as they are today, and the initial process of setting up a respective continental body lasted two years until AFC was finally founded in the course of the second Asian Games held in Manila. It took three meetings before AFC was formalised on 8 May 1954. Amongst its 12 founding member associations were Afghanistan, Burma, Taiwan, Hong Kong, India, Indonesia, Japan, South Korea, Pakistan, the Philippines and Vietnam.

A leading figure in initiating the inauguration of AFC was John M. Cleland, a Scot representing the Philippines. He was not only the representative who put a lot of effort into uniting the member associations behind the idea of establishing a confederation, but he was also to a great degree involved in working on the first AFC constitution, as well as drafting the rules for the first Asian Cup which was to be held for the first time in 1956 in Hong Kong, thereby being the second-oldest continental competition of its kind (following in the footsteps of the Copa America which had been taking place ever since 1916). According to the constitution, the representatives holding official positions within AFC were elected for a four-year term from 1954 until 1958 (Asian Football Confederation, no year: 8).

At the FIFA Congress held in Berne on 21 June 1954, AFC was recognised as a confederation and henceforth received one seat on the FIFA Executive Committee. While John M. Cleland's long-lasting commitment to the newly founded body found its expression in his role as Vice-President from 1954 until 1970, the first four Presidents spanned a period of just four years in total. All of these Presidents were citizens of Hong Kong, with Man Kam Loh serving for half a year, followed by Kwok Chan who kept the post until 1956, before being succeeded by S. T. Louey for one year and finally N. C. Chan, who took over until May

1958. With Hong Kong playing a leading role in the first few years of AFC – the first General Secretary, Lee Wai Thong, also came from the British colony – the office was located in Hong Kong (Asian Football Confederation, 1999: 3).

The period of political instability of AFC came to an end when the first Prime Minister of Malaya, Tunku Abdul Rahman Putra Al-Haj, was elected President during the first AFC Congress held in Tokyo on 29 May 1958 in the course of the Asian Games. The new President, who had also been elected President of the FAM in 1951, was committed to the progress of Asian football against the background of his belief in sports serving as an effective instrument in order to enhance social unity amongst different races, cultures and religions. For the purpose of celebrating the independence of Malaya, Tunku Abdul Rahman had initiated the construction of the Merdeka Stadium, which came to be the venue of the first Merdeka Tournament in 1957. The tournament emerged as a popular competition for the years to come, attracting international squads from all across the continent. Operating with a limited budget of HK$ 3300, the AFC President made the decision to organise a charity match after the Merdeka Tournament in 1958, which was played between Malaya and a selection of other international players. The fundraising event was aimed at increasing AFC's financial resources in order to finance more competitions and tournaments. Due to the President's conviction that Asian football had to improve its potential through drawing upon the development of youth and grassroots football, AFC inaugurated respective youth tournaments such as the Asian Youth Under-19 Championship, which took place for the first time in Malaysia in April 1959. Furthermore, under the aegis of Tunku Abdul Rahman the Asian Club Championship came into existence in 1967. Also, the first Asian Women's Championship was held in Hong Kong in 1975, albeit under the umbrella of the Asian Ladies Football Confederation, which was later incorporated into AFC in the early 1980s. Meanwhile AFC's office had moved to Penang, Malaysia, in 1965, with Koe Ewe Teik being the new General Secretary. He was succeeded by Dato Teoh Chye Hin nine years later and the office was relocated to Ipoh before finally moving to Kuala Lumpur in 1978 (Olympic Council Malaysia, no year-a; Asian Football Confederation, no year: 8; 1999: 3–4; 1997: 139; 2004: 24ff., 34ff., 104f.).

In the course of the presidency of Tunku Abdul Rahman, membership of AFC grew steadily, while in some cases conflicts emerged with significant implications for the political leadership. When AFC decided to expel Taiwan and Israel in the 1970s, a major conflict emerged with FIFA. What has been described in official AFC publications as a 'stormy period [...] with many unexpected events taking place' in the years between 1974 and 1978 culminated in Tunku Abdul Rahman resigning from his post in 1977 and being replaced by the Iranian Kamiz Atabi, who acted as President until the next Congress in December 1978 (Asian Football Confederation, no year: 8; 1999: 4; Moon and Burns, 1985: 2). In the elections, the member associations voted for the President of the FAM, Tan Sri Hamzah Abu Samah, to become the new President to represent AFC. Also the Malaysian Peter Velappan, who had previously gained administrative experience as Assistant Secretary for the FAM, was appointed General Secretary. This change in personnel marked the beginning of a new era, considering that Hamzah maintained the post until 1994 and Velappan until 2007. In almost 30 years serving as General Secretary, the latter established himself as one of, if not, the major football administrator giving Asian football a face and leading it into the new millennium.

While running a tiny secretariat with a handful of staff and sharing offices with the Olympic Council Malaysia throughout the 1980s and early 1990s, Velappan was highly influential with regard to formulating and executing new directives. Apart from extending the scope of responsible committees and installing new tournaments, such as the Asian Youth Under-16 Championship in 1984, the Asian Cup Winners' Cup in 1990 and the Super Cup in 1995, Velappan helped put more emphasis on the necessity of embracing the commercial value of football under consideration of the economic potential of the growing Asian market. After AFC had entered a contract with the sports marketing company International Sport and Leisure during the early 1980s, AFC enhanced its marketing activities in the 1990s through the establishment of a partnership with AFC Marketing Limited, today known as World Sport Group, according to which the latter gained the rights for marketing and broadcasting deals. Covering a four-year period, the first contract signed in 1993 was worth US$10 million, hence AFC could generate significantly more income than ever before and invested the benefits into its competitions and development policies (Asian Football Confederation, no year: 8; 1996: 2; 2004: 24, 36, 41; 1984: 13; Olympic Council Malaysia, no year-b; Hashim, 2011; Sugden and Tomlinson, 1998: 178).

In 1994, Tan Sri Hamzah decided to retire due to health reasons and, following the tradition of previous elections, another Malaysian, His Royal Highness Sultan Ahmad Shah, became the next President. Besides his functions as President of the FAM and the ASEAN Football Federation, the Sultan's role as the constitutional head of Pahang and the Head of Islam in the state required much of his presence so that he performed the AFC presidency in a more figurehead manner, thereby putting Peter Velappan effectively in charge of most strategic and operational matters. While the early 1990s saw AFC's constituency expanding to more than 40 member associations due to the dissolution of the Soviet Union, the second part of the decade and the early 2000s were devoted to laying the foundations for setting up AFC as a major actor in the regional and global spheres of football business. Not were only a website and a new official magazine launched, but a whole new corporate identity campaign was initiated including a new logo and the AFC's slogan 'The Future is Asia'. The objective of putting Asia on the map and drawing more attention to the continent was supported by the hosting of the World Cup in South Korea and Japan in 2002.

In organisational terms, the most significant event though was the inauguration of AFC House in Kuala Lumpur on 17 May 2000, which was built with the financial support of FIFA, while Sultan Ahmad Shah had proved to be an important agent in negotiating with the Malaysian government on the estate AFC was to receive. Whereas during the mid-1990s the staff of AFC counted not more than 12 employees – a surprisingly small number given the size of the continent and the number of member associations – the new headquarters set the basis for a large-scale expansion with new departments and professionally trained staff and personnel. In a sense, these developments therefore reflected the growing significance of the governing body and its new self-understanding, namely 'that Asia will soar in the world football scene in the near future' through 'the unity, solidarity and strength of AFC while setting the trend of professionalism for the new millennium within Asian football' (Hashim, 2011; Velappan, 2011; Church, 2011; incl. quote Asian Football Confederation, 2002: 4, 7; 2004: 26–27, 38; 1998: 7–8).

In 2002 Sultan Ahmad Shah decided to step down and at the Congress in August 2002 the Qatari Mohamed Bin Hammam, who had been the President of the Qatar Football Association from 1992 until 1996 and a member of the AFC Executive Committee, succeeded in running for the office of AFC President. The election of the first Arab as highest representative of AFC indicated a significant change, given that all previously elected Presidents had either come from Hong Kong or Malaysia. It reflected a shift of power relations against the backdrop of some of the Middle Eastern countries focusing on boosting their football cultures. Also, Bin Hammam was a member of the FIFA Executive Committee and the FIFA Goal Bureau, thereby exercising a notable degree of political power at the global level. Pointing to the improvements that had been made in Qatar in relation to developing grassroots and youth football, Bin Hammam dedicated his presidency to taking drastic measures in order to further develop and professionalise Asian football. Accordingly, the Asian Club Championship, the Cup Winners' Cup and the Super Cup were merged into the AFC Champions League, which came to be the new flagship club tournament of AFC. Alongside the Asian Cup, the President's Cup and the Challenge Cup were introduced as second-tier club and national team competitions aimed at increasing the number of matches at a competitive level for less-developed footballing nations. In terms of football development policies, Bin Hammam was also the brainchild behind the establishment of the long-term Vision Asia programme, which was launched in September 2002 aimed at employing a hands-on strategy in order to help the less-developed countries. Moreover, the AFC Professional Leagues Project was put into being, aimed at defining new criteria for fully professionalising the big leagues. Due to the expanding activities of AFC, new departments were installed and the number of employees therefore increased rapidly.

Although Bin Hammam was re-elected twice, his leadership style and policy strategies found several critics. Amongst them was Peter Velappan, who had decided to retire from his post as General Secretary in 2007 in order to hand it over to Paul Mony Samuel, who criticised Bin Hammam for being arrogant and for having centralising tendencies within the organisation, as well as for marginalising the big member associations. However, Bin Hammam was supported by most representatives from Asia when he became the first Asian football official ever to decide to run for the FIFA Presidency in 2011. While some observers opined that he had serious chances of representing Asia at the very top of global football politics, his attempt and also Bin Hammam's active tenure as AFC President ended prematurely when it was revealed that he had been accused of paying bribes to officials of the Caribbean Football Union, in the course of which the FIFA Ethics Committee provisionally suspended Bin Hammam from all future football activities. As a consequence, Zhang Jilong from China, a longtime AFC official and Vice-President since 2002, was chosen to function as AFC Acting President. Zhang Jilong expressed his view that the incident, as well as the match-fixing scandals in China and South Korea, had tarnished the image of Asian football and that it would therefore be necessary to implement reforms in order to fight corruption, improve administrative standards and create transparency. In July 2011 the AFC Executive Committee took initial measures by putting up an ad-hoc committee responsible for assessing and evaluating the current challenges deriving from corruption, match fixing and bribery (Weinberg, 2012: 538–539; Velappan, 2003; 2011; AFP, 2008; Asian Football Confederation, 2007: 8–10; 2011; 2011a; 2011b; FIFA, 2011; Dorsey, 2011).

Despite the rather negative headlines that have subsequently surrounded AFC and its decision-makers in particular, as well as the structural problems and challenges that remain, one has to keep in mind the overall development of AFC throughout the last six decades and how it has turned into a professionally run governing body responsible for regulating and developing the game for 47 member associations, all coming from very diverse sub-regions with different resources and capabilities. In the same vein, AFC General Secretary Alex Soosay (2011) noted that the development of AFC 'has been remarkable ever since it was founded in 1954' and that 'from being a small confederation it has gone to become a major factor in the development and administration of the game'. In retrospect, the reasons for this development are that on the one hand 'Asian football in general has received significantly more recognition and acceptance, especially in the last 20 years in commercial, economic and footballing terms', and that on the other hand 'these achievements would not have been possible without the role of the member associations which themselves have fostered the game technically and economically'.

References

AFP. (2008). 'Hammam lashes out at Velappan.' Retrieved from http://thestar.com.my/sports/story.asp?file=/2008/11/27/sports/2654518&sec=sports on 19 February 2011.
Asian Football Confederation. (no year). *Achievements 1979–1989 and Vison 90s*, Kuala Lumpur.
Asian Football Confederation. (1984). *AFC XIth Congress 1984.*
Asian Football Confederation. (1996). *AFC XVIIth Congress 1996: Standing committee reports and activities 1994–1996.*
Asian Football Confederation. (1997). *The A–Z of Asian football 97–98*, Hong Kong: EFP International Ltd.
Asian Football Confederation. (1998). *AFC XVIIIth Congress 1998: Reports and activities (1994–1998).*
Asian Football Confederation. (1999). *Directory 1999.*

Asian Football Confederation. (2002). *AFC XXth Congress 2002: Reports and activities (2000–2002)*.

Asian Football Confederation. (2004). *The power of dreams: 50 years of the Asian Football Confederation*, Kuala Lumpur.

Asian Football Confederation. (2007). *AFC XXIInd Congress 2007: AFC Activity Report 2004–2007*.

Asian Football Confederation. (2011). 'Message of unity from AFC ExCo.' Retrieved from www.the-afc.com/en/news/35880-message-of-unity-from-afc-exco on 18 September 2011.

Asian Football Confederation. (2011a). 'Statement by AFC Sr VP Zhang Jilong.' Retrieved from www.the-afc.com/en/news/35811-statement-by-afc-senior-vice-president-zhang-jilong on 18 September 2011.

Asian Football Confederation. (2011b). 'Zhang Jilong assumes office.' Retrieved from www.the-afc.com/en/member-association-news/east-asia-news/35282-zhang-jilong-assumes-office on 13 June 2011.

Church, Michael (sport journalist and Editor of *AFC Magazine* 1995–2001) interviewed by B. Weinberg, 26 September 2011, telephone interview.

Dorsey, J. M. (2011). 'Asian Football Federation moves to dismantle Bin Hammam's legacy.' Retrieved from http://mideastsoccer.blogspot.com/2011/08/asian-football-federation-moves-to.html on 17 September 2011.

FIFA. (2011). 'Ethikkommission sperrt Fussballoffizielle.' Retrieved from http://de.fifa.com/aboutfifa/organisation/bodies/news/newsid=1479070/index.html?cid=rssfeed&att= on 18 September 2011.

Hashim, Rizal (Malaysian sports journalist) interviewed by B. Weinberg, 27 September 2011, Kuala Lumpur.

Moon, P. and Burns, P. (1985). *The Asia–Oceania soccer handbook*. Oamaru: Paul Moon.

Olympic Council Malaysia. (no year-a). 'Hall of fame: Tunku Abdul Rahman.' Retrieved from www.olympic.org.my/museum/hof/ind/tarn.htm on 1 December 2011.

Olympic Council Malaysia. (no year-b). 'Hall of Fame: Dato' Peter Vellapan.' Retrieved from www.olympic.org.my/museum/hof/ind/dpv.htm on 1 December 2011.

Soosay, Alex (AFC General Secretary) interviewed by B. Weinberg, 28 September 2011, Kuala Lumpur.

Sugden, J. and Tomlinson, A. (1998). *FIFA and the contest for world football: Who rules the peoples' game?* Cambridge: Polity Press.

Velappan, P., (2003). 'Message from the General Secretary.' *Football Asia: The official magazine of the Asian Football Confederation*, February, p. 3.

Velappan, Peter (AFC General Secretary 1978–2007) interviewed by B. Weinberg, 11 October 2011, Kuala Lumpur.

5 AFC's relations and constellations

AFC and FIFA

According to the FIFA Statutes published in August 2011, AFC is recognised by the world governing body as one of six confederations. Besides being obliged to work with FIFA in relation to organising, regulating and developing the game, and hence 'comply with and enforce compliance with the Statutes, regulations and decisions of FIFA', the confederations consult with FIFA on membership and management-related issues concerning the national associations. Furthermore, the confederations have the right to appoint members to FIFA's Executive Committee (FIFA, 2011a: 17–19); AFC holds four seats including one FIFA Vice-President (FIFA, 2012a). Specifically regarding the confederations' Statutes and structure, FIFA requires approval before amendments, as proposed by their member associations or the Executive Committee, are made (Asian Football Confederation, 2010a: 7). Moreover, AFC can file requests to the FIFA Disciplinary Committee, for instance in relation to disciplinary matters occurring during World Cup qualifiers (Asian Football Confederation, 2011a). Finally, the confederations receive considerable financial support from FIFA. In 2009, AFC received a development grant of US$2.5 million. Also, FIFA owned assets worth US$2,519,321 million (Asian Football Confederation, 2011b: 7).

Due to the dual membership of national associations with both FIFA and AFC, it is important to follow a mutual approach. Regarding so-called 'governance' matters, AFC would therefore work together with FIFA and negotiate with the member association in question as well as the governmental actors that may be part of the national sport system. The process would thus include consultation with FIFA before sending a mission team in order to assess the situation together with the local actors. Since AFC as the responsible continental governing body has closer links with the member associations, it would then report to FIFA and make respective recommendations. Due to the fact that if a member association is suspended by FIFA it is also automatically suspended from AFC, the latter is generally interested in finding ways to prevent or correct cases of government interference or deviance in order to maintain the unified system as prescribed by the FIFA Statutes (Johnson, 2011). Recent cases in which FIFA and AFC engaged include Bahrain, Indonesia, Kuwait and Brunei (FIFA, 2011b; Asian Football Confederation, 2011c, 2009a, 2011d).

From a historical perspective, AFC was recognised by FIFA on 21 June 1954 shortly after it had been founded in May. The roles of the confederations in general remained a debatable topic within FIFA for the years to come. In 1957, issues arose with regard to the status of associations that were members of FIFA but not of AFC and vice versa. FIFA therefore asked AFC to follow the instructions and invite the member associations (China, Cyprus, Saudi Arabia, Iraq, Laos, Jordan, Lebanon, Syria and Turkey) to join AFC as well as to cancel the memberships of those associations (Nepal and North Borneo) that were not affiliated with FIFA unless they did so (FIFA, 1957: 10–11). In fact, not until the FIFA Congress 1961 in London did the confederations became incorporated into the FIFA Statutes, according to which they were responsible for organising tournaments that included clubs of more than two member associations as well as youth tournaments and working together with FIFA in relation to all aspects regarding international or other competitions. It was stressed that all members of a confederation should be affiliated with FIFA in order to secure the basic right of electing representatives to the FIFA Executive Committee (FIFA, 1962: 414). On 22 October 1963 the FIFA Executive Committee approved a proposal to be presented to the national associations and the confederations in which the raison d'être of the confederations was defined as not only contributing to the quality and quantity of regulating matches and tournaments throughout the continents, but also as being in a unique position to enhance the spirit of the game through promoting its popularity. For this purpose, FIFA decided that confederations could apply for grants in order to finance youth and amateur tournaments. Moreover, in 1964 the President of FIFA, Sir Stanley Rous, wrote a column in the FIFA official bulletin solely devoted to the roles of the confederations, in which he emphasised the importance of the confederations for developing the game and installing new competition formats, as well as cooperating with FIFA, because the confederations were in a better and closer position to the clubs and countries in the regions (FIFA, 1964: 536–539).

Despite the positive rhetoric produced by FIFA regarding the significance of the confederations, AFC officials were not pleased with how their organisation was represented in world football. At the 1961 FIFA Congress the Asian delegates expressed their desire to increase the number of seats on the FIFA Executive Committee from one to three, including the Vice-President and two other members. When FIFA decided to make a compromise offer by which AFC would receive two seats in total, the President of the FAS, Soh Ghee Soon, argued that a confederation with more than 30 members should be adequately represented in the bodies of FIFA. This view was substantiated by the fact that at the time UEFA, with its 32 countries, was represented through two Vice-Presidents and four members, CAF with nine countries by one Vice-President and one member, and CONMEBOL with 11 countries by one Vice-President and two members. Anyhow, the request was rejected leading to a furious reaction by Lim Kee Siong, who said: 'This is not justice and fair play. It is discrimination.' In fact, AFC had to wait another seven years until FIFA amended its Statutes at the Congress in Guadalajara, Mexico, and provided Asia with three seats on its Executive Committee (incl. quote Asian Football Confederation, 2004: 45–46).

In the 1970s, Asian football officials grouped behind the campaign of João Havelange to become FIFA President and initiate a change within the mostly Eurocentric organisation and thus give more voice to non-European member associations. At the Congress in 1974, Peter Velappan as FAM official was amongst the delegates who voted for Havelange and thereby laid the basis for a new era in FIFA. Under the new presidency, the commercialisation and global development of the game were put on the agenda. What had only vaguely taken shape in the 1960s was accelerated in the course of the late 1970s. AFC administrators highly appreciated the efforts Havelange made through installing further coaching schools, youth academies and other educational schemes. The financial and logistic support in the technical areas was also complemented by sharing administrative knowledge and expertise with the Asian member associations (Sugden and Tomlinson, 2003: 59; 1998: 174–176; Eisenberg, 2004: 13–14; 2007: 232–234; Samuel, 2008).

In contrast to the fruitful collaboration in the technical and administrative sectors, the 1970s and early 1980s were characterised by a major conflict between FIFA and AFC on how to handle the memberships of China, Taiwan and Israel. After AFC had decided to expel Taiwan and Israel while affiliating with the People's Republic of China, FIFA threatened to suspend AFC and impose drastic sanctions. AFC, however, was not impressed and the decision was ratified at the 1974 AFC Congress. In relation to Taiwan it was argued that only one member association could represent China, while the Arab countries had succeeded in lobbying for their cause to expel Israel. FIFA opined that this move constituted a breach of its Statutes. What complicated the matter was that the People's Republic of China had not been a FIFA member since 1958 while refusing to accept the 'Two Chinas' policy, and it therefore demanded that Taiwan had to be expelled. FIFA, however, did not make any concessions but rather continued putting pressure on AFC in order to readmit Taiwan. AFC in turn did not back down and reaffirmed its decision at the AFC Congress in Hong Kong in 1977. The relations between FIFA and AFC had become so strained that some Asian members even considered creating an alternative global governing body, together with some African constituents. At the FIFA Congress in Buenos Aires in 1978 the issue was handed over to the Executive Committee, which then decided to readmit the People's Republic of China while requiring Taiwan to change its name into Chinese Taipei. At the 1980 FIFA Congress the decision was approved, but it took another year until Taiwan agreed to modify the name of its association. While Israel had to compete with Oceanian countries to qualify for the World Cup until it finally affiliated with UEFA in the 1990s, it was not before 1989 that the Chinese Taipei Football Association finally re-entered the AFC (Homberg, 2006: 81ff.; Ritter, no year: 84).

Another issue that emerged during the mid-1970s was the establishment of the Asian Ladies Football Confederation (ALFC) and assessment of its authority status within the football system. After the ALFC had been formed in 1968, it organised four Asian Cup Women's Football tournaments between 1975 and 1981. The ALFC involved countries such as Hong Kong, Malaysia, Singapore, India and Taiwan, all of which were determined to promote women's football not only in Asia but also throughout the world. Thus in 1975 the ALFC requested to be recognised by FIFA which, however, opined that the ALFC had to become incorporated into AFC. At first neither AFC nor most of the men's national associations were at all interested in women's football and therefore declined any requests. When the ALFC General Secretary Charles D. Pereira informed his FIFA counterpart, Helmut Käser, about plans of organising a Women's World Cup in 1977 and the idea of founding a world governing body for women's football, FIFA became very concerned and stated that it would be impossible to recognise or affiliate with any independent organisation, as proposed by the ALFC. Due to the overall growth of women's football across the globe and under consideration of the ambitions formulated by women's football officials, FIFA increased the pressure on AFC and the national associations to take control of the regulation of the women's game. Nonetheless, some associations, such as the FAM, maintained their view that they did not want to encourage women's football. Meanwhile, FIFA approved the hosting of a Women's Football Invitation Tournament in Taipei in 1981 before sponsoring another tournament in 1983, both of which have been interpreted as the attempt to replace the ALFC Asian Cup, not least because AFC had made a respective suggestion. In the course of several trips FIFA delegates made to Asia during the early 1980s, the matter was discussed but it remained unresolved until November 1983, when AFC could express its confidence that the ALFC's resistance to women's football being integrated into the men's governing bodies had diminished. Notwithstanding, the entire process of incorporation was not completed before 1986 (Eisenberg *et al.*, 2004: 190–191; Asian Football Confederation, 1984: 13–14; Williams, 2007: 96–98).

Whereas the latter half of the 1980s and the 1990s were dominated by the expansion of collaborative efforts in the technical sector, highlighted through the establishment of the Futuro Coaching Programme, as well as in organisational and administrative terms, in the form of regular meetings that were held at the General Secretary level, new sources of conflict emerged towards the turn of the millennium regarding the role of the confederations and the individuals involved. Some confederations demanded a stronger voice and respective reforms, particularly with regard to organising World Cup qualifying matches as well as preselecting candidates for hosting the World Cup tournament. And at the individual level Asian officials such as South Korea's Chung Mong-Joon, who had been a member of the FIFA Executive Committee since 1994, tried to gain more influence through promoting the cause of Asian football and the World Cup 2002.

Meanwhile, Joseph Blatter successfully ran for the post as FIFA President in 1998. Although AFC as a whole had refrained from making any prior commitments to either Blatter or his competitor Lennart Johansson, the Qatari Mohamed Bin Hammam, who had joined the FIFA Executive Committee in 1996, turned out to be strong supporter of Blatter. In this context one has to differentiate on the one hand between the common interests of the Asian member associations and their officials, expressed in the mutual objective of representing and promoting AFC as a united actor, and on the other hand the potential for internal conflict deriving from the heterogeneity of the AFC's constituency and the political ambitions of its individual officials. Bin Hammam, for instance, despite his positive relations with Blatter, expressed his discontent with the decision to co-host the World Cup 2002, arguing that it posed an insult to Asia as a whole because it would imply that no Asian country was capable of hosting the tournament independently. Furthermore, he was one of the delegates who walked out at the 1999 Extraordinary FIFA Congress in Los Angeles in protest of neither being awarded a higher number of places for the World Cup nor another seat on the FIFA Executive Committee. In fact, Bin Hammam resigned from all FIFA Standing Committees he had been appointed to and AFC threatened to boycott the World Cup 2002, arguing that a confederation like CONMEBOL with ten members could not have four (and a possible fifth via a play-off match against another confederation's nation) and AFC only four places in the World Cup. Based on this argument, AFC even banned the South Americans from attending any of their events or meetings.

The action did not have any immediate impact on the outcome of the Congress apart from infuriating Blatter, who viewed the walk out as an insult. Then AFC President Sultan Ahmad Shah, however, contributed to restoring the good relationship with FIFA through several discussions with Blatter and, because the Asian member associations amounted to almost one-fourth of FIFA's constituency, all actors were interested in finding a solution. As a matter of fact, the negotiations resulted in giving four and half World Cup seats to Asia with UEFA making the concession of offering a play-off match between an Asian and a European team (FIFA, 2000: 15; Asian Football Confederation, 1996: 7; 1998a: 1; 2000: 7; 2002a: 17–18; Sugden and Tomlinson, 2003: 26, 59, 116, 155, 156; 1998: 163–164).

A recent case regarding the political relationship between FIFA and AFC and its individual representatives can be found in Bin Hammam's decision to run for the FIFA Presidency in 2011. Although some members of the media had speculated that one day he might succeed Sepp Blatter, Bin Hammam was also considered to be one of Blatter's strongest supporters. In fact, he frequently expressed his support for Blatter in speeches and other statements. Things changed when Bin Hammam met with Chung Mong-Joon in Seoul in February 2010 and mentioned that it would soon be time for an Asian FIFA President. However, after it was speculated that Blatter and Chung had had a major dispute during a meeting in Zurich in August 2010, Bin Hammam stated that he would not be interested in running and rather preferred focusing on his role as AFC President while supporting Blatter for a new mandate. A few days later Chung, who had previously mentioned that FIFA could do with some more competition, also retreated from the idea of opposing Blatter. The whole situation took a different direction when in March 2011 Bin Hammam finally did decide to challenge the incumbent. In his campaign the AFC President, who had only been re-elected in January 2011, made the case for changing decision-making structures in FIFA and making it more transparent. He proposed replacing the Executive Committee by a FIFA board containing 41 members and making their remunerations public, decentralising the administrative activities through delegating more competences to the confederations, including all stakeholders in the decision-making processes, installing a transparency committee, introducing a limited term for the presidency, making the voting for World Cup public, and providing more financial assistance for the member associations.

Against the backdrop of these ambitious objectives, Bin Hammam received support from most of the Asian member associations whereas UEFA backed Blatter's candidature. But before it came to the contest FIFA Executive Committee member Chuck Blazer accused Bin Hammam of having paid bribes to members of the Caribbean Football Union, upon which the Qatari withdrew his candidature and gave way to Blatter. Following the decision by the FIFA Ethics Committee, according to which Bin Hammam was banned from all football-related activities, Zhang Jilong became AFC Acting President as well as a FIFA Executive Committee member and also Chairman of the FIFA Organising Committee for the Olympic Football Tournaments.

After Asian football politics had received another setback due to allegations of corruption and misuse of FIFA funds against the Thai FIFA Executive Committee member Worawi Makudi – which, however, were found to be false by a FIFA investigation – AFC officials put forward a new outlook and direction by the end of 2011. Under the leadership of Jilong, who had explicitly called for drastic reforms in Asian football, and in light of the commitment by the new FIFA Vice-President and Deputy Chairman of the FIFA Development Committee, Jordan's Prince Ali Bin Al Hussein, who had always been a strong supporter of Blatter and defeated Chung in January 2011 in the AFC elections for the FIFA Executive Committee, not only were issues of transparency, corruption and match fixing put on top of the agenda but also religion-related matters such as the hijab rule. While Jilong and Prince Ali have stressed the importance of collaborating and maintaining good relations with FIFA, Blatter praised the achievements of AFC through having united the member associations in difficult times (Weinberg, 2012: 540–541; Warshaw, 2011a; Bin Hammam, 2011a, 2011b; BBC Sport, 2011; Asian Football Confederation, 2011e, 2011f, 2012a, 2011g; Mackay, 2011; Corbett, 2011).

Besides having worked together in order to organise the international match calendar and regulate the global transfer market, collaborative efforts in the technical area have been expanded and intensified throughout the last few years. At times, difficulties have occurred due to different understandings and approaches towards implementing development policies. In this sense the roles and interests of the political leaders have also affected the development work. Whereas there has always been a consensus on the importance of coordinating all initiatives and programmes, whatever the political problems at the top level, FIFA felt that under Bin Hammam AFC was not entirely supportive of FIFA's programmes, for instance as for selecting course participants or defining coaching license requirements. AFC in turn has perceived FIFA's position at times as not suitable for the challenges in Asia. Lately, though, a control mechanism was installed with regard to ensuring selection criteria and requirements for coaching courses. Annual coordination meetings now serve as a forum for synchronising programmes. Also, FIFA has appreciated that AFC has become more open-minded and determined to represent a unified Asia, despite the political ambitions of individual officials. Due to the fact that AFC, in contrast to confederations such as UEFA, has established itself as a very proactive organisation in relation to technical and development matters, the need to work together has been underlined by the mutual objective of coordinating the Vision Asia and the new FIFA Performance programmes in order to share knowledge, avoid duplications, assess the necessities of particular member associations, and therefore complement existing development programmes such as the Financial Assistance Programme, the Goal Programme and the Futuro Courses. Moreover, AFC has recently joined the annual meetings for all Asian General Secretaries organised by FIFA, in the course of which AFC itself has invited them to Kuala Lumpur in order to enhance collaborative measures with regard to education, administration and regulation (FIFA, 2004: 26; Borja, 2012; Al Sabah, 2011; Flath, 2012; Garamendi, 2011).

AFC and other confederations

The AFC Statutes do not include any codified bilateral relationships with other confederations, apart from the objective to foster friendly relations with all, while the overall functions of the confederations are commonly defined by the FIFA Statutes (Asian Football Confederation, 2011h: 9). As all confederations are represented in FIFA's bodies, AFC engages with UEFA, CAF, CONMEBOL, CONCACAF and the OFC, primarily at the global level. The political relations between the confederations become particularly relevant with regard to the division of power within FIFA, the election and allocation of FIFA posts, financial support schemes, and the decision to allocate FIFA tournaments. On a bilateral basis, meanwhile, some confederations have made agreements with each other in order to collaborate and exchange expertise regarding competition formats or technical development matters.

During the 1970s, AFC and CAF strengthened their bilateral relationship against the background of their perception to create a stronger voice within FIFA for non-European member associations. Therefore they set-up an Afro-Asia Consultative Committee with its first meeting held in Paris in November 1977. Henceforth the committee met at least once per year, mainly in the course of the confederations' Congresses or nations' tournaments. Furthermore, AFC and CAF established two intercontinental competitions, namely the biannual Afro-Asia Nations' Cup from 1985 and the annual Afro-Asia Club Championship from 1987. They also both decided to exchange referees for their tournaments as well as to foster technical cooperation. The Afro-Asian competitions as well as the good relationship with CAF, however, came to an end in 2000 following the decision by Asian delegates to vote for Germany instead of South Africa as host country for the World Cup 2006. This in turn affected the FIFA presidential elections in 2002 in Seoul, where Asian member associations still had CAF's reaction in mind and thus voted for Blatter and against his African challenger Issa Hayatou.

The Afro-Asian series was then replaced by the so-called AFC/OFC Challenge Cup, in which the best team of Asia met the best one of Oceania. The biannual competition took place in 2001 and 2003. While at the club level FIFA had successfully installed the FIFA Club World Cup, including the champions from all confederations and thereby making a possible resumption of the Afro-Asia Club Championship actually unnecessary, CAF and AFC decided to resume their mutual competition formats. The renamed national team tournament, the AFC Asia/Africa Cup, however, was played only once in 2007 and the club competition was never relaunched. Also, the ambitious plan of installing tournaments for youth and women's teams never materialised, nor did the AFC's objective to establish the same sort of cooperation with CONCACAF and the OFC. An agreement with UEFA on installing annual matches between the respective Champions League winners also failed to be put in practice (Asian Football Confederation, no year: 93; 1996: 31; 2002a: 10, 27; 2007a; 2007b: 13–14).

Although historically AFC's relations with UEFA and CONMEBOL have been affected by the perception of the latter's strong representation in FIFA's organs, in recent years AFC and UEFA have embarked upon solid cooperation concerning technical development matters. This again exemplifies the need to differentiate between the political level with its individual interests, and the administrative level where staff work on a rather more issue-oriented basis. However, the probability that both spheres intersect and that daily operations are interrelated with the political situation is high, while there may be cases, especially in the technical area, which are not as politicised as others. After UEFA and AFC had strongly supported the cause of strengthening the roles of the confederations, and in light of UEFA's compromise offer for a play-off match for the World Cup 2002, the early 2000s were characterised by friendly relations resulting in the establishment of the first AFC Professional Coaching Diploma Course in cooperation with UEFA. Accordingly, one course module took place in Europe and UEFA provided assistance with regard to organising and implementing these. Furthermore, AFC consulted with UEFA on technical, administrative and organisational matters with regard to the installation of the Vision Asia programme and the new AFC Champions League, which was intriguing inasmuch as Bin Hammam expressed his wish that the competition competes with its European counterpart within ten years. Amongst the relations with all other confederations, AFC's exchanges and collaborations with UEFA have been considered to be the most important due to the perception that UEFA has set the ultimate benchmark for successfully running and promoting its competitions. Therefore UEFA officials have been frequently invited to Kuala Lumpur to participate in workshops or function as experts in seminars and conferences. In fact, then AFC Director of Competitions Suzuki Tokuaki stated that the mutual cooperation could perhaps one day lead to the installation of common competitions such as an Euro-Asia league, not least because the Asian market had caught the increasing interest of UEFA. While UEFA did not support Bin Hammam in his campaign for the FIFA Presidency, UEFA Vice-President Marios Lefkaritis and Senior Advisor to the UEFA President William Gaillard visited AFC in 2012 in order to discuss further perspectives of signing a memorandum of understanding and collaborating in the technical and educational areas (Asian Football Confederation, 2002a: 9; 2012b; 2012c; Flath, 2012; Duerden, 2009; Soosay, 2011; Tokuaki, 2011; Bisson, no year).

AFC and the regional federations

The regional federations – ASEAN Football Federation, East Asian Football Federation, South Asian Football Federation and West Asian Football Federation – are neither formally recognised by FIFA nor by AFC. As a matter of fact, there is no mention at all of these bodies in the AFC and FIFA Statutes. However, 42 member associations of AFC are affiliated with the regional federations. (See Table 5.1 for a list of member associations for each federation.) The status of these governing bodies is problematic inasmuch as they are not covered by the football system as defined by FIFA and AFC, according to which the national associations can become a member of a confederation but are not allowed to maintain relations with entities that are not recognised by FIFA or AFC. Also, member associations of AFC are prohibited to 'form themselves into regional associations or federations without the consent of FIFA and the AFC' (FIFA, 2011a: 17ff.; Asian Football Confederation, 2011h: 15). This means by implication that the member associations of AFC enjoy the informal consent of FIFA and AFC to be part of the regional federations. In fact, the AFF Constitution, for instance, requires its members to be affiliated with FIFA, while defining the duties of its General Secretary as acting as a link between the AFF, FIFA and AFC (ASEAN Football Federation, 2011). Thus it will be interesting to see which roles these federations may play in future. In fact, there are already several points in which FIFA, AFC and the regional federations intersect, for instance with regard to officials holding posts in several bodies and mobilising respective votes, or in relation to football development measures and know-how exchange, as well as commercial matters (Garamendi, 2011; Chon, 2011).

The oldest federation is the AFF. In 1982 an informal meeting, which took place between the meetings of the AFC Executive Committee, was held in Bangkok. Amongst the participants was AFC General Secretary Peter Velappan. The idea to found a regional federation for southeast Asia was based on the recognition that the establishment of regional tournaments could enhance the quality of the game and help it catch up with the more advanced east Asian countries. Two years later, six member associations (Brunei, Indonesia, Malaysia, Singapore, Philippines and Thailand) attended the first official meeting of

Table 5.1 Regional football federations and member associations

Member association	Year of affiliation
ASEAN Football Federation	
Football Association of Indonesia	1984
Football Association of Singapore	1984
Philippine Football Federation	1984
The Football Association of Thailand	1984
Football Association of Malaysia	1984
National Football Association of Brunei Darussalam	1984
Myanmar Football Federation	1996
Football Federation of Cambodia	1996
Vietnam Football Federation	1996
Lao Football Federation	1996
Federacao de Futebol Timor Loro-Se	2005
East Asian Football Federation	
Associacao de Futebol de Macau – China	2002
Chinese Football Association	2002
Chinese Taipei Football Association	2002
Football Association of The Democratic People's Republic of Korea	2002
Guam Football Association	2002
Japan Football Association	2002
Korea Football Association	2002
Mongolian Football Federation	2002
The Hong Kong Football Association Ltd	2002
Northern Mariana Islands Football Association (associate member)	2008
South Asian Football Federation	
All India Football Federation	1997
Pakistan Football Federation	1997
Football Federation of Sri Lanka	1997
All Nepal Football Association	1997
Bangladesh Football Federation	1997
Football Association of Maldives	1997
Bhutan Football Federation	2000
Afghanistan Football Federation	2005
West Asian Football Federation	
Federation Libanaise de Football Association	2001
Football Federation Islamic Republic of Iran	2001
Iraqi Football Association	2001
Jordan Football Association	2001
Palestinian Football Association	2001
Syrian Football Association	2001
Qatar Football Association	2009
United Arab Emirates Football Association	2009
Yemen Football Association	2009
Bahrain Football Association	2010
Kuwait Football Association	2010
Oman Football Association	2010
Saudi Arabia Football Federation	2010

Sources: ASEAN Football Federation, no year; East Asian Football Federation, 2011; Wikipedia, 2012; Asian Football Central, 2011; West Asian Football Federation, no year.

the AFF in Jakarta. Its office was located in Kuala Lumpur. What followed was the installation of an ASEAN Club Championship between 1984 and 1989, which in fact decided which team would advance to the AFC's Asian Club Championships. After a period in which the AFF lay idle, it was revived in the mid-1990s with the aim to streamline administrative and educational activities amongst the member associations. Assisted by the then general secretaries of the FAM and the FAT respectively, Paul Mony Samuel and Worawi Makudi, and with the cooperation of the AML, the AFF launched its new flagship tournament for national teams called the Tiger Cup from 1996. Then AFC General Secretary Peter Velappan appreciated this development and said that such regional tournaments could help the smaller nations. Taking place every two years, it was later renamed into Suzuki Cup. In addition, youth tournaments, the Women's Championship as well as the AFF Futsal Championship were inaugurated. Counting 11 member associations, the AFF's presidency has been held by the former AFC President Sultan Ahamd Shah ever since 2011, while the general secretariat is led by the former AFC General Secretary Paul Mony Samuel. In recent times the AFF and World Sport Group have explored the possibility of establishing an ASEAN Super League (ASEAN Football Federation, no year; Velappan, 2000a; *New Straits Times*, 2011; *The Straits Times*, 2011).

The East Asian Football Federation, meanwhile, was established in May 2002 against the backdrop of the World Cup in South Korea and Japan. Aimed at developing east Asian football, improving communication amongst the national associations and promoting peace in the region, it consisted of eight founding member associations. Later, Hong Kong and the Northern Mariana Islands joined the EAFF, too. The major competition organised by the EAFF is the East Asian Football Championship. Further tournaments are the EAFF Women's Championship, the Futsal Championship and various youth competitions. Despite of not having any formal linkages with AFC, several EAFF Executive Committee meetings have been either chaired or attended by AFC officials such as Bin Hammam, Makudi or Samuel. Also, the EAFF has discussed matters covered by AFC, such as the admission of the Northern Marianna Islands or Australia. In the latter case the EAFF issued a letter to AFC complaining about the imbalance of the number of member associations in the four AFC-defined regions (West, South and Central, ASEAN, East),[1] given the fact that the number of seats on the AFC Executive Committee are allotted on a regional basis and that AFC had decided to locate Australia within southeast Asia. Other aspects discussed by the EAFF related to changes in the AFC Statutes as well as adopting the AFC's requirements for specific competition formats. Another link is the EAFF's representatives, for they include people like Zhang Jilong as Honorary Vice-President and Kohzo Tashima as Vice-President, both of whom hold AFC posts. Last but not least, the EAFF also expressed its support for Bin Hammam to become FIFA President in 2011, which AFC reported on its website (East Asian Football Federation, 2011b; 2005a: 9; 2005b; 2007; 2008; 2009; 2010; 2011a; 2011c; Asian Football Confederation, 2011j).

Due to a scarcity of available sources, not much can be said about the South Asian Football Federation. It was founded in 1997 by representatives of six member associations, thereby making it the second-oldest regional federation in Asia. Later on, Bhutan and Afghanistan joined the SAFF, in 2000 and 2005 respectively. The main tournament organised by the SAFF is the biannual SAFF Championship, which is also known as the South Asian Football Federation Cup, and took place for the first time in 1997 after it had emerged from its predecessors, the South Asian Association of Regional Cooperation Gold Cup in 1993 and the South Asian Gold Cup in 1995. The SAFF's current President is Kazi Salahuddin, who is also the President of the Bangladesh Football Federation. His forerunner was the Nepalese Ganesh Tapa, who is a current member of the AFC Executive Committee. In April 2011 an AFC delegation including Bin Hammam, as well as some FIFA officials, visited New Delhi, where Bin Hammam promised more technical and financial support for SAFF member associations. In turn, it seemed that SAFF supported his aspirations to become FIFA President (Asian Football Central, 2011; *Republica*, 2009; Amin-ul Islam, 2011; Bisson, 2011a).

The West Asian Football Federation as an organisation was established in 2000 under the auspices of Prince Ali Bin Al Hussein, who is its current President as well as a FIFA Vice-President and an AFC Executive Committee member. The WAFF first consisted of only six member associations but expanded to 13 by 2010. Its foundation was based on the idea to unite the region's national associations and expand the possibilities of competing and developing the game at an international level. The general objective is the enhancement of technical, educational and organisational cooperation amongst the countries in west Asia. The major tournament the WAFF organises is the biannual WAFF Championship. Further tournaments have been put in place for women and youth, as well as Futsal. In his role as WAFF President, Prince Ali has been determined to strengthen the regional federation's status vis-à-vis AFC, as well as to foster collaboration amongst the regions and with AFC. Besides the fact that the WAFF has worked together with AFC with regard to competition organisation, for which the WAFF has requested AFC match officials, the AFC Acting President Zhang Jilong attended the WAFF Congress in 2011, where he stressed that due to the size of the Asian continent regional federations such as the WAFF play a key role in developing the game and securing unity and solidarity amongst the national associations (West Asian Football Federation, no year; Zureikat, 2011; Asian Football Confederation, 2011k).

AFC and its member associations

In its Activity Report for the years 2007 until 2008, the Asian Football Confederation (2009b: 10) has described its relations with its member associations as follows:

> Football's voice and interests are carried over to AFC House in Kuala Lumpur through AFC's 46 member associations. [...] Mutual trust, respect and utmost cooperation are the principles governing the relationship

between AFC and its constituents. [...] AFC firmly believes that Asian football's future lies in the hands of its Member Associations, who are the real drivers of the game, and the confederation is committed to working closely with them.

As a matter of fact, AFC was founded by national associations and thus, by the very nature of its existence, it is first of all an organisation which is based on its constituents, namely the member associations (see Table 5.2 for a full list). With the historical aim of developing, organising and regulating the game in Asia, AFC has been and continues to be a confederation that not only incorporates its members and represents their interests, but also provides a forum for discussion and debate. At the same time, AFC itself constitutes an independent body with professional staff and respective objectives and ambitions, not least due to the organisational development and evolution of AFC, which has contributed to its status vis-à-vis the member associations.

According to the AFC Statutes, its primary objectives are to promote friendly relations with and amongst member associations as well as to protect their interests and resolve disputes. Generally, a member association should come from Asia. If an association from another continent without any affiliation would like to join AFC, FIFA would have to accept this. There are three types of membership: ordinary, provisional and associate. While an ordinary member has to be decided upon at the Congress, the latter two may be awarded by the AFC Executive Committee for a limited period of time. An associate member is a national association which is not affiliated with FIFA (e.g. Northern Mariana Islands). Both provisional and associate members are not allowed to vote at the AFC Congress or to hold any office. For logistical, representative and organisational purposes, AFC has divided its ordinary members into four different zones (Asian Football Confederation, 2011h: 9, 11–13).

Amongst the rights of each ordinary member are the rights to vote at the Congress, to propose points for the agenda and amendments to the Statutes, to nominate candidates for all elected bodies, to be informed about the activities of AFC via its official bodies, as well as to participate in the AFC's competitions and other activities or programmes. The obligations of the member associations include to fully comply with AFC and FIFA Statutes, to guarantee the election of their decision-making organs, to participate in AFC competitions, to pay a membership fee of US$500 annually, to request permission to play an international match and pay levies of at least US$500 when playing an international 'A' men's match, to respect the Laws of the Game, to accept that any dispute is handled under the jurisdiction of the arbitration tribunal responsible as defined by AFC and FIFA, to communicate any statutory amendments, to be in charge of regulating all leagues and clubs in their territory, as well as to maintain their independence from third parties. The latter aspect is particularly relevant in case governmental authorities interfere with the member associations' affairs, for this would constitute a reason to be suspended from FIFA and AFC. Furthermore, it is noteworthy that a member association can resign from AFC and that AFC can

Table 5.2 AFC member associations

Member association	Year of affiliation with AFC	Year of affiliation with FIFA	Year of foundation
Afghanistan Football Federation	1954	1948	1933
Football Association of The Democratic People's Republic of Korea	1954	1958	1945
The Hong Kong Football Association Ltd	1954	1954	1914
All India Football Federation	1954	1948	1937
Football Association of Indonesia	1954	1952	1930
Japan Football Association	1954	1929	1921
Korea Football Association	1954	1948	1933
Myanmar Football Federation	1954	1948	1947
Pakistan Football Federation	1954	1948	1947
Philippine Football Federation	1954	1930	1907
Football Association of Singapore	1954	1952	1892
Chinese Taipei Football Association	1955	1954	1936
Football Federation of Cambodia	1957	1954	1933
The Football Association of Thailand	1957	1925	1916
Football Federation Islamic Republic of Iran	1958	1948	1920
Football Association of Malaysia	1958	1954	1933
Saudi Arabia Football Federation	1959	1956	1956
Kuwait Football Association	1962	1964	1952
Yemen Football Association	1962	1980	1962
Federation Libanaise de Football Association	1964	1936	1933
Vietnam Football Federation	1964	1964	1962
Football Federation of Sri Lanka	1969	1952	1939
Syrian Football Association	1969	1946	1936
Bahrain Football Association	1970	1968	1957
National Football Association of Brunei Darussalam	1970	1972	1959
Iraqi Football Association	1971	1950	1948
All Nepal Football Association	1971	1972	1951
Qatar Football Association	1972	1976	1972

Bangladesh Football Federation	1974	1976	1972
Chinese Football Association	1974	1931	1924
Jordan Football Association	1974	1956	1949
United Arab Emirates Football Association	1974	1974	1971
Associaçao de Futebol de Macau – China	1976	1978	1939
Oman Football Association	1979	1980	1978
Lao Football Federation	1980	1952	1951
Football Association of Maldives	1986	1986	1982
Guam Football Association	1992	1996	1975
Bhutan Football Federation	1993	2000	1983
Football Federation of Kyrgyz Republic	1994	1994	1992
Tajikistan Football Federation	1994	1994	1936
Football Association of Turkmenistan	1994	1994	1992
Uzbekistan Football Federation	1994	1994	1946
Mongolian Football Federation	1998	1998	1959
Palestinian Football Association	1998	1998	1928
Federacao de Futebol Timor Loro-Se	2002	2005	2002
Football Federation Australia	2005	1963	1961
Northern Mariana Islands Football Association (associate member)	2009	–	2005

Sources: Asian Football Confederation, 2011i; FIFA, 2012b.

actually be dissolved upon a resolution passed by at least three-quarters of the member associations that are eligible to vote (Asian Football Confederation, 2011h: 14–17, 43, 46; Kuan, 2011; All India Football Federation, 2011).

As for disciplinary matters regarding specific matches (including club matches), AFC would inform the responsible member association, which in turn can appeal against the decision made by the AFC Disciplinary Committee. Examples include cases of supporters' misconduct or players' ineligibility (Asian Football Confederation, 2011l, 2011m, 2011n; Saudi Arabian Football Association, 2011). AFC also works together with and draws upon the advice of the member associations in order to optimise the AFC Disciplinary Code, for example via the creation of a catalogue on potential infringements (Asian Football Confederation, 2011o; Vietnam Football Federation, 2011; Japan Football Association, 2011).

With regard to proposals made by the member associations to modify the AFC Statutes, various aspects have been raised. In 2009, for instance, the KFA suggested incorporating an article on good conduct and behaviour, while Guam proposed to abandon the age limit for candidates to be elected to the Executive Committee and Saudi Arabia thought of reducing the maximum term of the presidency to merely four years in order to prevent the need for campaigning and collecting votes. All of these proposals, however, were not accepted (Asian Football Confederation, 2009c, 2010a).

Meanwhile, examples of member associations that have been suspended – primarily because of interference through state authorities but also because of match fixing – include Iran, Indonesia, Brunei, Thailand and Kuwait. Israel was expelled from AFC due to pressure of the Arab countries, while Kazakhstan resigned from the confederation in 2001 in order to join UEFA after it had been a member of AFC since 1994 (Asian Football Confederation, 1998b: 7; Football Asia, 2001). The most recent ordinary member is the Football Federation Australia (FFA), founded in 2004, which joined AFC by the beginning of 2006 after its affiliation had been ratified at the AFC Extraordinary Congress in 2005. Already during the 1970s the FFA's predecessor, the Australian Soccer Federation, which had been founded in 1961 amid heavy conflicts with the Australia Soccer Association regarding control of the game, had sought affiliation with AFC. The request was, however, rejected and Australia maintained its affiliation with the OFC. The reasons for most AFC member associations to oppose Australia's continuous wish to join AFC was that, first, Australia's strength might reduce the chances for other countries to qualify for the World Cup, and second, that the political context – with Australia having a rather difficult role in the Asia-Pacific region as well as the country's eurocentric culture – did not contribute to a better image. In fact, Malaysia was one of the countries putting quite some effort into preventing the Australians from becoming part of AFC. The resistance, however, did not last forever and in 2005 it was finally decided that the FFA could soon send its teams to AFC competitions. On 5 January 2011 it was announced that Australia will host the Asian Cup in 2015 (Asian Football Confederation, 2007c: 10–11; 2011p;

Murray and Hay, 2006; Mosley, 2006: 59ff.; Bubalo, 2005: 3; Football Federation Australia, 2011).[2]

In relation to tournament organisation and hosting, educational and technical assistance and aspects of professionalisation, AFC has worked together with the member associations through providing knowledge and expertise, financial resources as well as personnel, and implementation schemes. AFC has issued regulations for hosting associations, including aspects of organisation and logistics, marketing and media, as well an official protocol for tournaments and commercial guidelines. During the lead-time of major tournaments such as the Asian Cup, the AFC Organising Committee would consult frequently with the hosting association in order to ensure a successful organisational procedure. As for bilateral development policies, AFC has made efforts through the installation of Vision Asia, the Financial Assistance Programme (until 2012 known as AID27), as well as the General Secretaries Induction Programme. The latter is aimed at improving communication and working rapport with and amongst the associations. Meanwhile, with regard to enhancing the professional structures of Asian football, AFC has defined a set of club licensing criteria that have to be fulfilled in order to compete in specific AFC competitions. This initiative had been brought about in cooperation with the JFA, which could draw upon its experience from the J-League. Therefore, AFC asked its member associations to ensure that all affiliated professional clubs participating in AFC competitions meet the criteria by 2013. In this context, AFC created the Professional Football Development Taskforce which visits various countries in order to help the associations to meet the requirements. And especially for the development of Indian football, AFC has installed an Ad-hoc Committee for Indian Professional Football (Asian Football Confederation, 2011q; 2010b: 14; 2011r; 2007a; 2008a; 2011s; 2011t; 2012d; 2010c; 2012d; 2011u; 2011v).

While AFC has always been determined to stress the unity and solidarity with and amongst its member associations, divisions and power blocs have been unavoidable for the simple reason that each national association is interested in possibly maximising its benefit from its membership or maintaining its sphere of influence vis-à-vis AFC. Apart from the cultural, political and economic diversity, which characterises the Asian continent and therefore contributes to the heterogeneity of the AFC's constituency, one also has to acknowledge the significance of power/political ambitions held by individual officials. In fact, various observers have referred to the creation of power blocs within AFC, particularly between the west and the east as well as the big and the small associations. Examples hereof include the shift of power during the 1990s, when the Middle Eastern officials gained more influence in the decision-making processes of the AFC bodies, not least resulting in the nomination of Lebanon for the Asian Cup 2000 and the election of Bin Hammam as AFC President in 2002. In turn, the CFA has followed the objective of securing a strong representation in AFC and subsequently in FIFA ever since the late 1970s. Also, the relationship between AFC and the member associations has been interpreted as at times tense, specifically with regard to AFC's push towards professionalisation, harmonising league structures and regulating competition formats throughout Asia. A prominent example is the conflict between AFC and the FAM in the course of the Asian Cup 2007, when the FAM, as one of the hosting associations of the tournament, arranged for a Manchester United tour match although it had signed an agreement with AFC according to which it was not to invite any foreign teams for the duration of the Asian Cup. However, under pressure from AFC and FIFA, the FAM cancelled the event.

Moreover, the opinion has been expressed that the role of the administration of AFC could be strengthened, given the perception that key decision-makers in the bodies of AFC, but also in the national associations, were supposedly too concerned with individual ambitions and not competent enough regarding football-specific issues, whereby the potential of a win–win situation for all actors, and especially the coaches and players, was reduced. Against this backdrop, the approach from the AFC administration has been to all the more embrace a pan-Asian perspective and employ diplomatic skills in order to overcome power/political divisions and discrepancies, in particular with regard to ensuring a unified image of AFC at the global level.

Not least, the case of Bin Hammam running for the FIFA Presidency, as well as the programmes initiated during his mandate as head of AFC, have been viewed with discord. On the one hand, a considerable number of Asian associations backed the idea of an Asian at the top of FIFA; on the other hand, the accusations of bribery, which tarnished the image of Asian football, as well as the perception that Bin Hammam's policies had intensified the political divisions between the associations, led to reactions voicing the need of introducing reforms and dissolving the links between power/political interests and AFC's development policies. At the same time, the incident initiated a discussion on a potential successor of Bin Hammam thereby, fuelling the debate on power blocs and partisanship (Sugden and Tomlinson, 1998: 162ff.; Tan and Bairner, 2010: 588–589; AFP, 2007; *The Star*, 2007; Asian Football Confederation, 2007b: 27; Careem, 2011; Velappan, 2011; Gautam, 2011; Church, 2011; Hashim, 2011; J-League employee, 2011; Fink, 2011; Soccerex, 2011a; Duerden, 2011).

AFC and the leagues, clubs and players

AFC is not directly affiliated with the leagues, clubs or players, but instead with their respective national associations. According to the Statutes of AFC, the clubs, leagues, regional associations or any other stakeholders have to be subordinate to as well as recognised by an AFC member association. Furthermore, the member associations are required to define the rights and duties of the clubs or stakeholders, as well as to approve their Statutes and regulations. It has to be ensured that the clubs' and stakeholders' decisions with regard to membership issues are made independently. Also, the member associations are responsible for preventing third-party control over more than one club or stakeholder by any natural or legal person, in case the integrity of a match or tournament might be endangered. Hence, the overall influence of AFC on the leagues and clubs, as well as the players, functions primarily via the member associations (Asian Football Confederation, 2011h: 18).

The fact, however, that AFC organises club competitions constitutes an immediate connection between it and the participating clubs. The AFC Executive Committee is responsible for defining a club licensing system, including criteria for clubs participating in AFC competitions and the requirements that have to be fulfilled by the licensing bodies (member association/league). Amongst the requirements are aspects regarding the organisation, technical standards, attendance, management, marketing and promotion, business scale, game operation, media, stadiums and the clubs. Depending on the type of competition – President's Cup, AFC Cup or AFC Champions League – the standards which have to be met vary. The minimum requirements in order to participate in the ACL season 2011/12 were a system of promotion and relegation as well as that the clubs were registered as a commercial entity. As for the number of foreign players allowed to play in the ACL and AFC Cup matches, AFC has introduced a '3+1' rule, according to which a team may field three non-Asian and one Asian foreigner. Meanwhile the teams competing in the AFC President's Cup may field two foreigners, while the quotas in the national leagues differ from country to country.

The requirements and club licensing criteria, as well as the establishment of the AFC Professional Football Development Taskforce and the implementation of financial analyses regarding the leagues and clubs, have indeed affected all Asian leagues and clubs because the natural interest of clubs lies in sporting success, not only at the national but also at the international level. In theory, all clubs should therefore have a genuine interest in participating in AFC club competitions, provided the reforms that have to be undertaken in order to meet the criteria and requirements are realistic. In fact, various leagues have expressed their ambition to improve their standards and AFC has provided its assistance and know-how, for example in Iran and Qatar. In this context, AFC has also responded to problems some leagues had with match fixing through sending special mission teams to countries such as South Korea (Asian Football Confederation, 2011h: 44–45; 2011t; 2009d; 2010d; 2011w: 37; 2011x: 37; 2011y: 35; 2011z; 2011aa; Kulkarni, no year; Jamali, 2011; Somerford and Kim, 2011; Soccerex, 2011b).

Another area in which AFC engages with the clubs as well their players concerns disciplinary matters. A prominent example is the case of the suspensions following an ACL semi-final between Suwon Samsung Bluewings and Al Sadd in 2011, in which it had come to a mass brawl involving players and coaches alike. Consequently, three players were sent off and AFC later suspended another two players and an assistant coach (Warshaw, 2011b). Of further relevance are of course the AFC Anti-Doping Regulations, which oblige all players to undergo doping tests. Having accepted the World Anti-Doping Code of 2009 and requiring all member associations to comply with FIFA's Anti-Doping Regulations, AFC aims at preserving the ethics of sport, ensuring the health of the players and equal chance through the implementation of mandatory tests. In the case of a positive investigation, players, clubs and associations can be respectively sanctioned and suspended. AFC maintains the right to save the attained data, including player whereabouts, as long as they are considered relevant and as required by the WADA Code (Asian Football Confederation, 2010e: 8–10, 24, 55).

It is noteworthy that no Asian counterparts to the European Club Association or the Association of European Professional Football Leagues exist, partly due to the fact that most Asian leagues are not as elaborately professionalised as in Europe. It is, however, only a question of time until similar organisations emerge, given the professionalisation schemes initiated by national associations as well as AFC. Meanwhile, players have organised themselves in various countries including India, Malaysia, Australia and Japan. In particular, in Australia the representation and support of players vis-à-vis the clubs, league and association through collective bargaining has been significant due to the fact that an organisation called Professional Footballers Australia, formerly known as the Australian Soccer Players' Association, has been in place since even before the Bosman ruling, specifically since 1993. Amongst other things, the PFA has achieved the establishment of a standard player contract and the abolition of the domestic transfer and compensation fee system.

Most importantly, the PFA has played a considerable part in the development of the world players' unions FIFPro Division Asia/Oceania, which consists of Australia, Japan, New Zealand, India, Indonesia and Malaysia. With the aim of representing and extending the voice of players throughout the continent, FIFPro Asia focuses on protecting and enforcing the legal rights of the players. For this purpose, the FIFPro Asia Legal Committee and Professional Football Committee have been established in order to provide advice on legal aspects to the board and to ensure the safety, career paths and interests of the players in the member countries. Accordingly, the division has consulted with member associations, such as the CFA, to strengthen and encourage the importance of organising a platform for the players. Also, the chairman of the division, Australia's Brendan Schwab, was appointed by the FIFA Executive Committee to the FIFA Dispute Resolution Chamber (DRC) for the period from 2009 until 2013. The DRC's objective is to provide players and clubs alike with a platform in order to foster effective resolution of disputes. Furthermore, the FIFPro Division Asia has signed memorandums of understanding with the Indonesian League Board and the ASEAN Football Federation, in 2008 and 2009 respectively, aimed at fostering dialogue and cooperation between the stakeholders with regard to equal representation and legal matters. In addition, the objective of negotiating a continental agreement with AFC has been brought forward through initial meetings with Bin Hammam as from December 2009. In fact, at the FIFPro Division Asia/Oceania Congress held in May 2012 in India, AFC expressed its interest in working closer together with the division and invited it to share its expertise against the backdrop of the confederation's professionalisation efforts. Also, then FIFA Vice-President and AFC Executive Committee member Prince Ali Bin Al Hussein asked Brendan Schwab to support his cause for changing the hijab rule.

Another recent indicator of AFC's interest in improving the conditions for players is the signing of a memorandum of understanding with three Centres of Sports Medicine Excellence in November 2011, according to which a selection of sports medicine fellows were to visit the centres in early 2012 in order to extend their knowledge and thereby improve the standards of health care service to players (PFA Malaysia, 2012; Professional Footballers Australia, 2012a; no year-a; no year-b; 2012b; FIFPro, no year; 2012a; 2012b; Asian Football Confederation, 2011ab).

AFC and governmental actors

The relationship between AFC and governmental organisations is based on the assumption by football governing bodies such as FIFA and AFC that they operate within an independent and autonomous sport infrastructure. Nonetheless, AFC does not exist in a legal vacuum, for it is registered as an association under Section 7 of the Societies Act of the Laws of Malaysia. Also, in some countries such as China, governmental involvement in sport affairs has been de facto relatively high. Moreover, in contrast to Europe, most Asian countries do not have a traditional club culture so that especially grassroots and youth football is controlled through ministries responsible for schools and physical education.

Another conduit is the involvement of politicians in sport governing bodies by holding official posts. The fifth AFC President, Tunku Abdul Rahman, was the Prime Minster of Malaysia during his time with AFC, while Sultan Ahmad Shah was politically active through his role as constitutional head of Pahang and the Head of Islam in the state. The latter's post and contacts did not turn out to be counterproductive for securing four acres of land from the Malaysian government in order to build the new AFC House in Kuala Lumpur. In fact, the so-called AFC Diamond of Asia Award, which honours a single or an ongoing contribution to Asian football and AFC, was awarded to the Malaysian Prime Minister Dato' Seri Najib Tun Razak at the annual AFC Awards ceremony in 2010. This followed a decision by the AFC Executive Committee in 2008 to allow other countries to bid to host the AFC headquarters, with the aim of improving the status of AFC vis-à-vis the hosting country and signing a memorandum of understanding covering aspects such as the legal status of AFC officials and employees, as well tax exemption options. The process, however, ended in Kuala Lumpur remaining the home of AFC (Malaysian Government, 1966: 13; Whittle, 2011; Asian Football Confederation, 2002b: 16; 2011ac; 2008b; AFP, 2008).

As has been outlined above, interference through state authorities in 'governance' matters is considered a violation that results in sanctions. In some cases the AFC regulations have actually required governments to change their laws, such as in Thailand where the national law prevented the President of the national association from holding the post longer than two years, whereas a period of four years is prescribed by AFC. In the context of the AFC's professionalisation schemes, according to which clubs participating in the ACL have to be registered as commercial entities, AFC would meet with governmental authorities in order to initiate necessary reforms. In the case of match-fixing and corruption scandals, such as in China and South Korea, AFC has attended meetings between national associations and state authorities including local police forces, but merely with an observer status; only in a few cases has AFC actively assisted the associations.

In the framework of its football development programme Vision Asia, AFC has frequently consulted with governmental actors such as ministries responsible for education and schools with the objective of discussing new ways for the development of youth and grassroots football, which has proven to be very important given the fact that the Vision Asia projects depend on collaboration with local actors who implement the suggestions made by AFC. In the context of the development initiatives, AFC representatives and in particular the AFC President would meet with state representatives in order to stress the importance of governmental support for the sport. Furthermore, in the course of the AFC's social responsibility programme Dream Asia, partnerships and contacts with governmental actors have been fostered and strengthened. The Midnight Football Project, for instance, in Kuala Lumpur has been organised together with the Malaysian National Unity and Integration Department. Another example is AFC's help for rebuilding sports facilities in conjunction with the government in Myanmar following a cyclone disaster in 2008. Most notably, AFC has entered a partnership with the UN Food and Agriculture Organisation in 2010, in the course of which AFC has made donations to several projects throughout Asia besides fostering awareness-raising through football for the fight against hunger (Johnson, 2011; Careem, 2011; Tokuaki, 2011; Sugihara, 2011; Flath, 2012; Asian Football Confederation, 2005a: 4; 2011ad; 2011ae; 2011af; 2011q; Yusa, 2011; FAO, 2010).

Finally, AFC relies upon and requires governmental support in the context of competition organisation and management. Therefore, potential bidding nations have to guarantee sufficient state support regarding security and infrastructure. Not least, after there had been concerns with the progress of completing the necessary facilities for the Asian Cup 2000 in Lebanon, which however was finally realised through the assistance of the Lebanese President General Emile Lahoud, AFC has increasingly put focus on the necessity of adequate collaboration with local governments. Thus, as stated in the regulations for the Asian Cup 2011, the organising association and the governmental authorities have to provide and implement a safety and security plan for all relevant locations (Tokuaki, 2011; Careem, 2007; Velappan, 2000; Asian Football Confederation, 2011x: 12).

AFC and commercial enterprises

After AFC sold its marketing rights to the International Sport and Leisure (ISL) in the 1980s, Peter Velappan and Seamus O'Brien decided to set-up AFC Marketing Limited (AML) in the early 1990s. Ever since 1993, the Singapore-based

AML, which was later renamed into the World Sport Group, has partnered with AFC in order to implement and promote the necessary marketing and media activities for major competitions. In parallel to the rapid development of AFC in the last two decades, the WSG has expanded its portfolio of sports rights and services including not only the major tournaments of AFC, the AFF, the WAFF and the SAFF, but also the Indian Premier League (cricket) and several golf events in Asia. Counting a total number of more than 150 employees, the WSG considers itself 'Asia's leading sports marketing, event management and media company' with a reach to more than 30 countries. Amongst its main business objectives is the commercial exploitation of sports events and sports rights in order to achieve the highest prossible benefits for its partners and their brands (Velappan, 2011; World Sport Group, no year; O'Brien, no year).

Constituting the main source of income for AFC, the partnership with the WSG has grown tremendously in terms of financial volume. While the first contract in 1993 made the WSG pay merely US$10 million for a cycle of four years, the sum increased to US$200 million in 2005 for a period of eight years. The latest contract was signed in 2009 for the period from 2013 until 2020 and provides AFC with a sum of at least US$1 billion in revenues. The agreement covers competitions such as the Asian Cups 2015 and 2019, the Champions League, the Asian qualifiers for the FIFA World Cups 2014 and 2018, the Women's Asian Cup, as well as the Futsal and youth competitions. Whereas according to prior contracts the WSG was the sole representative and fully responsible for arranging deals with potential sponsors or broadcasters, the new contract allows AFC to take part in tripartite meetings and thereby play a more active role in enhancing its brands and monitoring the sales processes. Amongst the commercial sponsors for the competitions are companies such as Emirates (2002–2013), Qtel (2009–2012), Epson (1997–2016), Makita (2005–2016), Toyota (2005–2016) as well as FXPro Financial Services (2011) (Asian Football Confederation, 2005b: 4; 2009e; Reuters, 2009; Saito, 2011; Kuan, 2011; Emirates Airline, 2002; Emirates, 2009; Glendinning, 2008; Soccerex, 2012; Bisson, 2011b).

AFC and other partners

In addition to those already mentioned in this chapter, AFC has relationships with several other actors which are part of the non-governmental and civil society sector. The German Football Association (Deutscher Fußball Bund or DFB) has been a partner in the context of technical cooperation initiatives ever since the turn of the millennium. It has provided AFC with support for appointing a technical director for grassroots and youth football in the years from 2003 until 2006, as well as sharing expertise with regard to football development strategies. Moreover, the DFB assisted with implementing a professional coaching licence module for Asian coaches in Germany in 2005.

Meanwhile, the English Football Association has shared its knowledge with AFC through producing a strategic development plan. Based on an FA/AFC

Technical Cooperation Agreement signed in 2002, the plan served as a blueprint and input for furthering grassroots and elite player development as well as coach education within the framework of the AFC's AID26, later AID27, programme. The Anglo-Asian dialogue and mutual work has prevailed until today. In 2012, AFC General Secretary Alex Soosay met with the Premier League's Chief Executive, Richard Scudamore, in order to improve the importance of the Premier League's development activities in Asia and how these can produce synergies with the AFC's policies, particularly with regard to grassroots football. In fact, Chelsea FC's then Chief Executive, Peter Kenyon, had already visited Kuala Lumpur in 2006 with the aim of discussing the objectives of the club in the Asian market and how it could get involved with AFC's activities. Hence a memorandum of understanding was signed according to which Chelsea FC provided technical and financial assistance for the AFC's Vision Asia projects in China from 2007. In 2011 the partnership was renewed and extended to other projects by means of providing instructors for coaching courses, management workshops and financial support (Asian Football Confederation, 2002b: 124; 2012f; 2006a; 2006b; 2011ag; 2011v; 2011ah; Flath, 2012; Bradbury, 2002; *FourFourTwo*, 2007; *Thaindian News*, 2008).

In the framework of social responsibility activities, AFC signed a memorandum of understanding in 2011 with a Qatar-based non-governmental organisation named Reach Out To Asia (ROTA), which is attached to the Qatar Foundation. ROTA is devoted to improving learning and life skills as well as increasing the literacy rate amongst children and youth in Asia. In the context of the AFC's Dream Asia Programme, ROTA and AFC have implemented for instance a football-based youth empowerment programme, which took place in the course of the Asian Cup 2011 in Qatar. In China, meanwhile, AFC has collaborated with the CFA and the China Soong Ching Ling Foundation based on a memorandum of understanding signed in 2011 and aimed at utilising football as a tool for empowering girls. Accordingly, a festival for underprivileged girls was organised during the AFC's Under-16 Women's Championship in Nanjing in order to promote life skills, self-confidence and a healthy lifestyle (Asian Football Confederation, 2011ai, 2011aj, 2011ak, 2011al, 2011am).

In 2012 AFC signed a collaboration agreement with the newly established Asian Football Development Project. The AFDP was founded and is chaired by FIFA Vice-President Prince Ali Bin Hussein, who created the non-governmental and non-profit organisation, which is based in Jordan, with the objective of providing assistance to Asian national and regional associations that are in need of help to improve grassroots and youth football, women's football and the social dimension of the game. As defined in the bilateral agreement, its further aim is to promote cooperation, transparency and professionalism in the administration and management of Asian football. The football-related NGO is particularly interested in working together with AFC in order to produce the highest possible degree of synergies in football development. Understanding itself as a conduit for the interests of all Asian football organisations, the AFDP has taken first steps through bringing together key experts in October 2011 in order to discuss the need to modify the hijab rule, whereupon the recommendations were communicated to the FIFA and AFC Executive Committees. Furthermore, the AFDP has launched an assistance scheme for which football organisations and even companies can apply. The thematic areas include youth programmes, women's football, social responsibility activities and the protection and innovation of the game. The first pilot project was launched in the Philippines, where the AFDP supported a grassroots festival for more than 1000 boys and girls. The future role of the AFDP as a proactive football development organisation will remain highly interesting, particularly due to the fact that it constitutes an entity beyond the traditional football system as established by FIFA and the confederations (Asian Football Confederation, 2012g; Asian Football Development Project, 2012; WeAreAsia, 2012).

Interim conclusions

AFC's relations are manifold and include various actors. The relations vary in significance, outreach and intensity. They are also intertwined with each other.

Most important is obviously AFC's relationship with FIFA. Recognised as one of six confederations, AFC has to comply with the requirements and standards defined by the global governing body. It benefits from FIFA by receiving financial and technical support. Both organisations collaborate with regard to organisational and statutory matters that affect the member associations, specifically when it comes to cases of governmental interference and respective measures to be taken. Until the late 1960s, FIFA was still very Eurocentric. This changed with the presidency of João Havelange, in the course of which the confederations received more support. However, conflicts in relation to how to treat the status and affiliation of countries such as China and Israel occurred, as well as how to deal with the ALFC. With Asian delegates constantly striving towards increasing their representation in FIFA's organs, aspects such as hosting the World Cup and defining the number of places for Asian countries have constituted sources of conflict and dissent. The recent case of Mohamed Bin Hammam running for the FIFA Presidency and being sanctioned by FIFA serves as a reflection. Finally, collaborative efforts in the technical sector have been expanded, as have mutual approaches with regard to transfer policies and fighting match fixing and corruption.

With regard to AFC's relations with other confederations, it has to be noted that bilateral relations have not been codified, but merely expressed in memorandums of understanding. In terms of political power, the relations with other confederations have always been characterised by attempts to protect and increase spheres of influence at the global level, specifically in FIFA's organs. Also, bidding for FIFA tournaments and securing financial support have had an impact on these relations. Nonetheless, collaborative efforts have led to the temporary installation of common competitions, such as with CAF. With UEFA, meanwhile, AFC has increased consultation and dialogue concerning competition and marketing policies as well as educational measures.

The relations AFC holds with the regional federations in Asia constitute a very interesting situation because the four federations are formally recognised by neither AFC nor by FIFA. However, 42 of the AFC's member associations are affiliated with the regional federations. Some of the AFC's representatives are also involved with the federations and Bin Hammam has even chaired meetings of the EAFF. In fact, Prince Ali Bin Al Hussein is President of the WAFF. Moreover, the significance of the regional tournaments organised under the auspices of the regional federations cannot be neglected when examining the future of Asian football.

The core function and very nature of AFC is essentially based on its constituents, namely the member associations. Having experienced a considerable increase in membership throughout the last decades, AFC incorporates its members and reflects their interests. In principle, the members have the right to dissolve AFC. Simultaneously, AFC has gained authority over its members: if national associations want to participate in the established system of football, membership of AFC is obligatory. While AFC can be considered a very proactive confederation, driving development and professionalisation throughout Asia, conflicts within AFC and between the members are not avoidable and have led to divisions and the emergence of power blocs. AFC tries to counter these through stressing the unity and solidarity amongst it members and by following a pan-Asian outlook.

AFC is not directly affiliated with the leagues, clubs and players, but mainly interacts with these through the member associations. The competitions organised by AFC, however, create a platform for immediate connections, finding their expression in requirements and regulations that the clubs, leagues and players have to comply with, e.g. transfer rules and the fielding of foreign players. In particular when it comes to disciplinary matters, AFC engages with these actors. It is also noteworthy with regard to collective bargaining processes that AFC has signed a memorandum of understanding with the Asian division of FIFPro.

The relations with governmental actors are primarily defined by the autonomy of the sport-political infrastructure. Therefore, cases of governmental interference form a crucial point of intersection. On the other hand, it is common that representatives who hold political positions in a country are members of AFC's organs. In the context of developing the game at the grassroots level, governments and ministries are the primary actors to engage with. This is for instance the case with respect to installing and implementing Vision Asia projects. Moreover, governments become relevant stakeholders when it comes to hosting tournaments and fulfilling the necessary requirements. In terms of finding solutions for match fixing and corruption, AFC has also discussed these issues with governmental actors.

Regarding the relations with commercial enterprises, AFC holds a partnership with the World Sport Group, formerly known as AFC Marketing Limited (AML). The partnership aims at implementing and promoting marketing and media activities for major competitions. For this purpose WSG commits to ensuring the commercial exploitation of the AFC's competitions and the respective rights. The partnership constitutes the major source of income for AFC, with the current contract generating a sum of US$1 billion in revenues. In this context, AFC has the right to participate in tripartite meetings with WSG and customers, as well as to enhance its brands and monitor the sales. Amongst the sponsors have been companies such as Emirates, Qtel and Epson.

Finally, AFC maintains relations with various civil society actors. The DFB and the English FA have partnered with AFC in order to provide assistance for technical and administrative matters, while Chelsea FC has engaged in supporting the Vision Asia projects in China. In the context of the social responsibility activities of AFC, a partnership with ROTA shall help contribute to the achievement of AFC's goals. Another collaborative effort has been institutionalised through a bilateral agreement with AFDP that aims at promoting cooperation, transparency and professionalism in the administration and management of Asian football.

Notes

1 The regional differentiation made by AFC does not correspond with the memberships of national associations in the regional federations. According to AFC, Iran, for instance, belongs to South and Central Asia, whereas it is a member of the West Asian Football Federation (Asian Football Confederation, 2011h: 13).
2 Further works on football in Australia have been produced by Hay (2010; 2011) and O'Hara (1994).

References

AFP. (2007). 'FA of Malaysia bow to pressure.' Retrieved from http://thestar.com.my/sports/story.asp?file=/2007/5/9/sports/17668045&sec=sports on 18 September 2011.
AFP. (2008). 'ACF boss offers olive branch to Malaysia.' Retrieved from http://football.thestar.com.my/starspecial/story.asp?file=/2008/11/19/footballeveryday/2581935&sec=FootballEveryDay on 18 September 2011.
Al Sabah, Hasan (AFC Deputy General Secretary and Director of Education) interviewed by B. Weinberg, 7 October 2011, Kuala Lumpur.
All India Football Federation. (2011). 'Letter to AFC requesting permission for international matches.'
Amin-ul Islam, M. (2011). 'SAFF ready to shed regional tag.' Retrieved from http://articles.timesofindia.indiatimes.com/2011-04-28/interviews/29482268_1_asian-football-federation-hammam-regional-tag on 18 September 2011.
ASEAN Football Federation. (2011). 'Constitution: Statutes of the Federation.' Retrieved from www.aseanfootball.org/?page_id=86 on 17 September 2011.
ASEAN Football Federation. (no year). 'About AFF: History of ASEAN Football Federation.' Retrieved from www.aseanfootball.org/?page_id=2 on 17 September 2011.
Asian Football Central. (2011). 'SAFF (South Asian Football Federation).' Retrieved from http://asianfc.com/video/saff-south-asian-football-federation on 20 June 2012.
Asian Football Confederation. (1984). *AFC XIth Congress 1984.*
Asian Football Confederation. (1996). *AFC XVIIth Congress 1996: Standing committee reports and activities 1994–1996.*

Asian Football Confederation. (1998a). *AFC News, 5 (No 5)*.
Asian Football Confederation. (1998b). *AFC News, 4 (No 10)*.
Asian Football Confederation. (2000). *AFC XIXth Congress 2000: Reports and activities (1998–2000)*.
Asian Football Confederation. (2002a). *AFC XXth Congress 2002: Reports and activities (2000–2002)*.
Asian Football Confederation. (2002b). *AFC XXth Congress 2002: Agenda*.
Asian Football Confederation. (2004). *The power of dreams: 50 years of the Asian Football Confederation*, Kuala Lumpur.
Asian Football Confederation. (2005a). *Newsletter*.
Asian Football Confederation. (2005b). *Newsletter*.
Asian Football Confederation. (2006a). *Newsletter*.
Asian Football Confederation. (2006b). *Newsletter*.
Asian Football Confederation. (2007a). 'Japan crush Egypt in AFC Asia-Africa Challenge Cup.' Retrieved from www.the-afc.com/en/member-association-news/east-asia-news/9318 on 19 June 2012.
Asian Football Confederation. (2007b). *AFC XXIInd Congress 2007: AFC XXIst Congress 2004 minutes and AFC Extraordinary Congress 2005 minutes*.
Asian Football Confederation. (2007c). *AFC XXIInd Congress 2007: AFC Activity Report 2004–2007*.
Asian Football Confederation. (2007a). *AFC Protocol regulations*.
Asian Football Confederation. (2007b). *AFC XXIInd Congress 2007: Agenda*.
Asian Football Confederation. (2008a). 'Commercial guidelines for hosting associations: AFC Asian Cup 2011 – final qualification round.' Retrieved from www.the-afc.com/uploads/Documents/common/cms/afc/AC2011QualifiersCOMMERCIAL.pdf on 5 October 2011.
Asian Football Confederation. (2008b). '*Criteria for AFC headquarters*.' Retrieved from www.the-afc.com/uploads/Documents/common/cms/afc/criteriahq.pdf on 5 October 2011.
Asian Football Confederation. (2009a). *Correspondence between FIFA, AFC and Kuwait FA on Suspension of KFA*.
Asian Football Confederation. (2009b). *AFC XXIIIrd Congress 2009: AFC Activity Report 2007–2008*.
Asian Football Confederation. (2009c). *AFC XXXIIIrd Congress 2009: Proposals and secondments for Congress agenda*.
Asian Football Confederation. (2009d). 'Strategy for the development of professional football in Asia' presentation.
Asian Football Confederation. (2009e). 'AFC, WSG renew landmark partnership.' Retrieved from http://www.the-afc.com/en/media-releases/26117-afc-wsg-renew-landmark-partnership on 10 November 2009
Asian Football Confederation. (2010a). *AFC Extraordinary Congress 2010: Proposed amendments to the Statutes*.
Asian Football Confederation. (2010b). *AFC Champions League 2010 manual*. Retrieved from http://images.the-afc.com/Documents/2010/acl2010manual.pdf on 5 October 2011.
Asian Football Confederation. (2010c). 'Ad-hoc Committee for Professional Clubs.' Retrieved from www.the-afc.com/en/tournaments/clubs/afc-champions-league/29884-afc-ad-hoc-committee-for-professional-clubs on 5 October 2011.
Asian Football Confederation. (2010d). *AFC club licensing manual*.
Asian Football Confederation. (2010e. *AFC Anti-Doping Regulations*.

Asian Football Confederation. (2011a). *Letter to participating member associations containing FIFA Disciplinary Committee decision on 17 August 2011 regarding the request filed by the AFC for the cancellation of cautions in the FIFA WC 2014 preliminary competition.*

Asian Football Confederation. (2011b). *AFC XXIVth Congress 2011: Financial report 2009.*

Asian Football Confederation. (2011c). 'Last chance for Indonesia.' Retrieved from www.the-afc.com/en/member-association-news/asean-news/35106-last-chance-for-indonesia on 31 May 2011.

Asian Football Confederation. (2011d). 'FIFA lift Brunei suspension.' Retrieved from www.the-afc.com/en/member-association-news/asean-news/35105-fifa-lift-brunei-suspension on 31 May 2011.

Asian Football Confederation. (2011e). 'Bin Hammam withdraws from FIFA Presidential election.' Retrieved from www.the-afc.com/en/member-association-news/west-asia-news/35080-bin-hammam-withdraws-from-fifa-presidential-election on 31 May 2011.

Asian Football Confederation. (2011f). 'Zhang Jilong assumes office.' Retrieved from www.the-afc.com/en/member-association-news/east-asia-news/35282-zhang-jilong-assumes-office on 13 June 2011.

Asian Football Confederation. (2011g). '*FIFA President hails AFC unity.*' Retrieved from www.the-afc.com/en/news/37256-fifa-President-hails-afc-unity on 2 December 2011.

Asian Football Confederation. (2011h). 'Statutes of the Asian Football Confederation: Regulations governing the application of the Statutes and standing orders of the Congress', 2011 edition. Retrieved from http://image.afcpresident.com/upload/library/AFC_Statutes_2011.pdf on 10 November 2010.

Asian Football Confederation. (2011i). 'Member associations.' Retrieved from www.the-afc.com/en/about-afc/member-associations on 19 September 2011.

Asian Football Confederation. (2011j). 'EAFF support for Bin Hammam.' Retrieved from www.the-afc.com/en/inside-afc/676-afc-news/34382-eaff-support-for-bin-hammam on 18 April 2011.

Asian Football Confederation. (2011k). 'Jilong praise for WAFF.' Retrieved from www.the-afc.com/en/member-association-news/west-asia-news/36514-jilong-praise-for-waff on 26 September 2011.

Asian Football Confederation. (2011l). 'Letter to Korea Football Association regarding disciplinary matters: AFC Champions League 2011 (knockout stage) Jeonbuk Hyundai Motors (KOR) vs. Cerezo Osaka (JPN) match on 27.9.11.'

Asian Football Confederation. (2011m). 'Letter to IR Iran Football Federation containing decision of the AFC Disciplinary Committee on 26 September 2011 regarding the case of Foolad Mobarekeh Sepahan.'

Asian Football Confederation. (2011n). 'Letter to IR Iran Football Federation regarding appeal against the decision passed by the AFC Disciplinary Committee on 26 September 2011.'

Asian Football Confederation. (2011o). 'Letter to member associations regarding catalogue on potential infringements under the AFC Disciplinary Code.'

Asian Football Confederation. (2011p). 'Australia to host 2015 AFC Asian Cup.' Retrieved from www.the-afc.com/en/news-centre/news/32028-australia-to-host-2015-afc-asian-cup on 27 June 2012.

Asian Football Confederation. (2011q). 'AFC celebrates World Food Day.' Retrieved from www.the-afc.com/en/programs-dream-asia-campaigns/campaigns-afc-fao/campaigns-afc-fao-news/36811-u-19-finalists-promote-578-message on 11 December 2011.

114 *AFC's relations and constellations*

Asian Football Confederation. (2011r). *AFC Asian Cup Qatar 2011: Marketing and media regulations.*

Asian Football Confederation. (2011s). 'Committee pledges to work closely with FFA.' Retrieved from www.the-afc.com/en/news/37227-committee-pledges-to-work-closely-with-ffa on 12 December 2011.

Asian Football Confederation. (2011t). *Criteria for participation in AFC club competitions 2011–2012.*

Asian Football Confederation. (2011u). 'Indonesia seeks AFC's expertise.' Retrieved from www.the-afc.com/en/news-centre/inside-afc/676-afc-news/36950-indonesia-seeks-afcs-expertise on 11 December 2011.

Asian Football Confederation. (2011v). 'Indian football assessed by committee.' Retrieved from www.the-afc.com/en/member-association-news/central-a-south-asia-news/37220-ad-hoc-committee-takes-stock-of-indian-football on 12 December 2011.

Asian Football Confederation. (2011w). *AFC Champions League competition regulations 2011.*

Asian Football Confederation. (2011x). *AFC Cup competition regulations 2011.*

Asian Football Confederation. (2011y). *AFC President's Cup competition regulations 2011.*

Asian Football Confederation. (2011z). *Financial analysis of Asian professional leagues and clubs.*

Asian Football Confederation. (2011aa). 'Qatari clubs set sights on ACL 2012.' Retrieved from www.the-afc.com/en/news-centre/inside-afc/676-afc-news/36590-qatari-clubs-set-sights-on-acl-2012 on 11 December 2011.

Asian Football Confederation. (2011ab). 'New milestone in players' healthcare.' Retrieved from www.the-afc.com/en/news-centre/inside-afc/676-afc-news/37243-new-milestone-in-players-healthcare on 12 December 2011.

Asian Football Confederation. (2011ac). 'AFC Diamond of Asia: HM Sultan Ahmad Shah.' Retrieved from www.the-afc.com/en/news-centre/news/37267-afc-diamond-of-asia-hm-sultan-ahmad-shah on 12 December 2011.

Asian Football Confederation. (2011ad). 'Jilong pays courtesy visit to PM of Mongolia.' Retrieved from www.the-afc.com/en/member-association-news/east-asia-news/37126-jilong-pays-courtesy-visit-to-pm-of-mongolia on 12 December 2011.

Asian Football Confederation. (2011ae). 'MF impresses M'sian unity dept.' Retrieved from www.the-afc.com/en/programs-midnight-football/programs-midnight-football-news/36403-mf-impresses-msian-unity-dept on 18 September 2011.

Asian Football Confederation. (2011af). 'AFC–FAO help 14,000 hungry people.' Retrieved from www.the-afc.com/en/programs-dream-asia-campaigns/campaigns-afc-fao/campaigns-afc-fao-news/36813-afc-fao-help-14000-hungry-people on 11 December 2011.

Asian Football Confederation. (2011ag). 'Chelsea make CSR touch through VA.' Retrieved from www.the-afc.com/en/football-development/vision-asia/33836-chelsea-make-csr-touch-through-va on 16 March 2011.

Asian Football Confederation. (2011aha). 'AFC, Chelsea FC renew partnership.' Retrieved from www.the-afc.com/en/football-development/vision-asia/33835-afc-chelsea-fc-renew-partnership on 6 July 2012.

Asian Football Confederation. (2011ai). 'AFC signs MoU with ROTA.' Retrieved from www.the-afc.com/en/social-responsibility-news/32034-afc-signs-mou-with-rota on 7 January 2011.

Asian Football Confederation. (2011aj). 'ROTA keen on future cooperation.' Retrieved from www.the-afc.com/en/dream-asia-award/programs-dreamasia-awards-news/37291-rota-keen-on-future-cooperation on 12 December 2011.

Asian Football Confederation. (2011ak). 'AFC Dream Asia award: Reach Out To Asia (ROTA).' Retrieved from www.the-afc.com/en/news-centre/news/37268-afc-dream-asia-award-reach-out-to-asia-rota on 12 December 2011.

Asian Football Confederation. (2011al). 'CFA, CSCLF vow continued cooperation.' Retrieved from www.the-afc.com/en/football-for-the-disabled/football-for-disabled-news/37020-cfa-csclf-vow-continued-cooperation on 11 December 2011.

Asian Football Confederation. (2011am). 'Fest for underprivileged girls kicks off.' Retrieved from www.the-afc.com/en/football-for-the-disabled/football-for-disabled-news/37014-fest-for-underprivileged-girls-kicks-off on 11 December 2011.

Asian Football Confederation. (2012a). 'Hijab issue: Jilong seeks FIFA support.' Retrieved from www.the-afc.com/en/news/38122-hijab-issue-jilong-seeks-fifa-support- on 13 June 2012.

Asian Football Confederation. (2012b). 'AFC gets UEFA support.' Retrieved from www.the-afc.com/en/news-centre/inside-afc/676-afc-news/37974-uefa-visit on 16 February 2012.

Asian Football Confederation. (2012c). 'AFC ExCo takes important decisions.' Retrieved from www.the-afc.com/en/news/38461-afc-exco-takes-important-decisions on 27 June 2012.

Asian Football Confederation. (2012d). 'Teamwork vital in AFC–MA relations.' Retrieved from www.the-afc.com/en/news-centre/inside-afc/676-afc-news/38182-teamwork-vital-in-afc-ma-relations on 27 June 2012.

Asian Football Confederation. (2012e). 'Fulfill club licensing criteria by 2013: AFC to MAs.' Retrieved from www.the-afc.com/en/events/elite-education-seminar-2012/37698-fulfill-club-licensing-criteria-by-2013-afc-to-mas on 16 February 2012.

Asian Football Confederation. (2012f). 'AFC, Premier League ties discussed.' Retrieved from www.the-afc.com/en/news/38769-afc-premier-league-ties-discussed on 5 July 2012.

Asian Football Confederation. (2012g). ;AFC, AFDP sign collaboration agreement.' Retrieved from www.the-afc.com/en/news/38462-afc-afdp-sign-collaboration-agreement- on 5 July 2012.

Asian Football Confederation. (no year). *Achievements 1979–1989 and Vison 90s*, Kuala Lumpur.

Asian Football Development Project. (2012). 'Call for proposals.' Retrieved from www.weareasia.com/en/the-afdp/Call-For-Proposal.pdf on 6 June 2012.

BBC Sport. (2011). 'Mohamed Bin Hammam calls for major change at Fifa.' Retrieved from http://news.bbc.co.uk/sport2/hi/football/9380806.stm on 30 January 2011.

Bin Hammam, M. (2011a). '*President Manifesto.*' Retrieved from http://image.afcpresident.com/upload/library/20110318_president_manifesto.pdf on 23 May 2011.

Bin Hammam, M. (2011b). Press conference declaring running for FIFA Presidency. Retrieved from http://image.afcpresident.com/upload/library/MBH_Press_conference_0318.pdf on 23 May 2011.

Bisson, M. (2011a). 'Bin Hammam pushes FIFA Presidential credentials in India: Blog boasts of achievements.' Retrieved from www.worldfootballinsider.com/Story.aspx?id=34278 on 15 April 2011.

Bisson, M. (2011b). 'New sponsor for Asia's Champions League; FIFA takes over Bosnia federation.' Retrieved from www.worldfootballinsider.com/Story.aspx?id=34271 on 15 April 2011.

Bisson, M. (no year). 'Bin Hammam's FIFA Presidential chances fade as UEFA leaders back Blatter.' Retrieved from www.worldfootballinsider.com/Story.aspx?id=34336 on 13 May 2011.

Borja, David (FIFA Development Manager) interviewed by B. Weinberg, 1 February 2012, Zurich.

Bradbury, J. (2002). 'FA present Asian football blueprint.' Retrieved from www.thefa.com/TheFA/InternationalRelations/NewsAndFeatures/2002/21693 on 1 December 2011.

Bubalo, A. (2005). *Football diplomacy.* Paper of the Lowy Institute for International Policy.

Careem, N. (2007). '2011 Asian Cup hopefuls must get government support.' Retrieved from http://sports.espn.go.com/espn/wire?section=soccer&id=2863245 on 5 October 2011.

Careem, Nazvi (sports journalist and AFC Media Director 2008–2011) interviewed by B. Weinberg, 1 October 2011, Kuala Lumpur.

Chon, Lim Kim (AFC Technical Director 1999–2004) interviewed by B. Weinberg, 3 October 2011, Kuala Lumpur.

Church, Michael (sports journalist and Editor of AFC Magazine 1995–2001) interviewed by B. Weinberg, 26 September 2011, telephone interview.

Corbett, J. (2011). 'Asian football chief Bin Hammam rues Chung's FIFA departure.' Retrieved from www.afcpresident.com/en/News/view_news.php?pageNum=2&menuNum=24&idx=153 on 19 September 2011.

Duerden, J. (2009). 'The Asian Football Confederation has laid down the law as it tries to raise the Asian Champions League's standing in the game.' Retrieved from www.guardian.co.uk/sport/blog/2009/mar/17/asian-champions-league-reorganisation on 5 October 2011.

Duerden, J. (2011). 'Exclusive: Bitter battle underway to replace Bin Hammam as Asian football boss.' Retrieved from www.worldfootballinsider.com/Story.aspx?id=34464 on 16 July 2011.

East Asian Football Federation. (2005a). *EAFF Official Guidebook.*

East Asian Football Federation. (2005b). '15th East Asian Football Federation Executive Committee meeting.' Retrieved from www.eaff.com/eanews/release/2002_2005/051214.html on 17 September 2011.

East Asian Football Federation. (2007). 'Regarding the agenda/decisions of the 23rd EAFF Executive Committee meeting.' Retrieved from www.eaff.com/eanews/release/2007/071218.html on 17 September 2011.

East Asian Football Federation. (2008). 'Regarding the agenda/decisions of the 24th EAFF Executive Committee meeting.' Retrieved from www.eaff.com/eanews/release/2008/080223.html on 17 September 2011.

East Asian Football Federation. (2009). 'Regarding the agenda/decisions of the 28th EAFF Executive Committee meeting.' Retrieved from www.eaff.com/eanews/release/2009/090227.html on 17 September 2011.

East Asian Football Federation. (2010). 'Regarding the agenda/decisions of the 30th EAFF Executive Committee meeting.' Retrieved from www.eaff.com/eanews/release/2010/101020.html on 18 September 2011.

East Asian Football Federation. (2011a). 'EAFF outline.' Retrieved from www.eaff.com/organisation/about.html on 17 September 2011.

East Asian Football Federation. (2011b). 'The history of East Asian football.' Retrieved from www.eaff.com/fanzone/history/index.html on 17 September 2011.

East Asian Football Federation. (2011c). 'EAFF statement regarding Bin Hammam's decision to run for FIFA Presidency.' Retrieved from www.eaff.com/eanews/release/2011/110415.html on 17 September 2011.

Eisenberg, C. (2004). 'Fußball als globales Phänomen: Historische Perspektiven.' *Aus Politik und Zeitgeschichte: Beilage zur Wochenzeitung 'Das Parlament'*, (26), 7–15.

Eisenberg, C. (2007). 'Der Weltfußballverband FIFA und das 20. Jahrhundert: Metamorphosen eines Prinzipienreiters.' In J. Mittag and J.-U. Nieland (eds), *Das Spiel mit dem Fußball: Interessen, Projektionen und Vereinnahmungen* (pp. 219–236). Essen: Klartext Verlag.

Eisenberg, C., Lanfranchi, P., Mason, T. and Wahl, A. (2004). *100 years of football: The FIFA centennial book*. London: Weidenfeld & Nicolson.

Emirates Airline. (2002). '*Emirates Airline joins the AFC commercial family*.' Press release. Retrieved from www.ameinfo.com/11432.html on 18 September 2011.

Emirates. (2009). 'Emirates remains committed to Asian football's future.' Retrieved from www.emirates.com/jp/japanese/about/news/news_detail.aspx?article=397114 on 10 November 2010.

FAO. (2010). 'AFC and FAO sign agreement to battle hunger.' Retrieved from www.fao.org/getinvolved/getinvolved-home/getinvolved-highlights/fr-detail-high/fr/item/42495/icode/ on 10 November 2010.

FIFA. (1957). *Agenda of the Executive Committee meeting in Moscow*.

FIFA. (1962). *Official bulletin, 37/38*.

FIFA. (1964). *Official bulletin, 41*.

FIFA. (2000). *Activities report April 1998–April 2000*.

FIFA. (2004). *Activity report April 2002–March 2004*.

FIFA. (2011a). '*New executive committee for Indonesia FA*.' Retrieved from www.fifa.com/aboutfifa/news/newsid=1473645/ on 5 October 2011.

FIFA. (2011b). *Statutes*.

FIFA. (2012a). 'FIFA Executive Bodies.' Retrieved from www.fifa.com/aboutfifa/organisation/bodies/excoandemergency/index.html on 6 June 2012.

FIFA. (2012b). 'FIFA's member associations.' Retrieved from www.fifa.com/aboutfifa/organisation/associations.html on 25 June 2012.

FIFPro. (no year). 'About division Asia/Oceania.' Retrieved from www.fifpro.org/division/index/2/show:details on 1 December 2011.

FIFPro. (2012a). 'Division Asia meets about India, Indonesia and AFC.' Retrieved from www.fifpro.org/news/news_details/1943 on 30 June 2012.

FIFPro. (2012b). 'FIFPro supports proposal for headscarf rule change.' Retrieved from www.fifpro.org/news/news_details/1857 on 30 June 2012.

Fink, Jesse (Australian sports journalist) interviewed by B. Weinberg, November 2011, email interview.

Flath, Thomas (AFC Director of Grassroots and Youth 2003–2006) interviewed by B. Weinberg, 13 February 2012, Mönchengladbach.

Football Asia. (2001). 'Kazakhstan resigns from AFC.' *Football Asia: The official magazine of the Asian Football Confederation*, July, p. 7.

Football Federation Australia. (2011). 'FFA submission to the "Australia in the Asian Century" white paper.'

FourFourTwo. (2007). 'Kenyon: AFC deal "crucial" for Chelsea profile.' Retrieved from http://au.fourfourtwo.com/news/46764,kenyon-afc-deal-crucial-for-chelsea-profile.aspx on 10 November 2010.

Garamendi, Domeka (FIFA Development Officer in Kuala Lumpur) interviewed by B. Weinberg, 23 September 2011, Kuala Lumpur.

Gautam, Biplav (AFC Vision Asia Officer 2007–2009) interviewed by B. Weinberg, 14 September 2011, telephone interview.
Glendinning, M. (2008). 'Qtel to sponsor 2011 AFC Asian Cup.' Retrieved from www.sportbusiness.com/news/168214/qtel-sponsor-2011-afc-asian-cup on 10 November 2010.
Hashim, Rizal (Malaysian sports journalist) interviewed by B. Weinberg, 27 September 2011, Kuala Lumpur.
Hay, R. (2010). 'A tale of two footballs: The origins of Australian football and association football revisited.' *Sport in Society*, *13*(6), 952–969.
Hay, R. (2011). 'Ethnicity, structure and globalization: An argument about association football in Australia, 1958–2010.' *Sport in Society*, *14*(6), 833–850.
Homberg, H. (2006). 'FIFA and the "Chinese Question", 1954–1980: An exercise of Statutes.' *Historical Social Research*, *31*(1), 69–87.
J-League employee (anonymous) interviewed by B. Weinberg, 4 October 2011, email interview.
Jamali, Omid (Head of FFIRI International Relations Department) interviewed by B. Weinberg, October 2011, email interview.
Japan Football Association. (2011). 'Letter to AFC regarding catalogue on potential infringements under the AFC Disciplinary Code.'
Johnson, James (AFC Director of International Relations and Development) interviewed by B. Weinberg, 11 October 2011, Kuala Lumpur.
Kuan, Bryan (Head of AFC Finance Department) interviewed by B. Weinberg, 30 September 2011, Kuala Lumpur.
Kulkarni, C. (no year). 'Professionalisation efforts gather steam.' Retrieved from www.the-afc.com/en/news/38437-professionalisation-efforts-gather-steam on 29 June 2012.
Mackay, D. (2011). 'Makudi cleared by FIFA over land deal.' Retrieved from www.insideworldfootball.biz/worldfootball/asia/10082-makudi-cleared-by-fifa-over-land-deal on 20 December 2011.
Malaysian Government. (1966). 'Laws of Malaysia. Act 335: Societies Act 1966.' Incorporating all amendments up to 1 January 2006. Retrieved from www.agc.gov.my/Akta/Vol.%207/Act%20335.pdf on 16 September 2011.
Mosley, P. (2006). 'Playing ball with Asia: Asian-Australian links through soccer.' In W. J. Murray and R. Hay (eds), *The world game Downunder* (pp. 53–74). Melbourne: Australian Society for Sports History.
Murray, B., and Hay, R. (2006). 'Introduction.' In W. J. Murray and R. Hay (eds), *The world game Downunder* (pp. 1–6). Melbourne: Australian Society for Sports History.
New Straits Times. (2011). 'Sultan Ahmad willing to serve.' Retrieved from www.nst.com.my/articles/39ajithelm/Article on 6 May 2012.
O'Brien, S. (no year). 'Chairman's message: World Sport Group.' Retrieved from www.worldsportgroup.com/about-us/message/ on 16 September 2011.
O'Hara, J. (ed.). (1994). *Ethnicity and soccer in Australia* (A.S.S.H. Studies in Sports History: No. 10).
PFA Malaysia. (2012). 'Welcome to PFA Malaysia.' Retrieved from www.pfamalaysia.com.my/ on 30 June 2012.
Professional Footballers Australia. (2012a). 'PFA History.' Retrieved from www.pfa.net.au/index.php?id=37 on 30 June 2012.
Professional Footballers Australia. (2012b). 'FIFA Dispute Resolution Chamber.' Retrieved from www.pfa.net.au/index.php?id=68 on 30 June 2012.
Professional Footballers Australia. (no year-a). 'FIFPro Asia and the legal rights of players.' Retrieved from www.pfa.net.au/index.php?id=114 on 1 December 2011.

Professional Footballers Australia. (no year-b). '*The ongoing development of FIFPro Asia.*' Retrieved from www.pfa.net.au/index.php?id=113 on 1 December 2011.

Republica. (2009). 'Ganesh Thapa no longer SAFF President.' Retrieved from http://archives.myrepublica.com/portal/index.php?action=news_details&news_id=10404 on 21 June 2012.

Reuters. (2009). 'AFC's rights deal with WSG to run until 2020.' Retrieved from http://thestar.com.my/sports/story.asp?file=/2009/11/25/sports/5179127 on 18 September 2011.

Ritter, M. (no year). *Der Wettbewerb um die Fußballweltmeisterschaft 2002.* Ruhr-Universität Bochum.v

Saito, Satoshi (Head of AFC Marketing Department) interviewed by B. Weinberg, 10 October 2011, Kuala Lumpur.

Samuel, P. M. (AFC General Secretary 2007–2009) interviewed by B. Weinberg, 5 August 2008, Kuala Lumpur.

Saudi Arabian Football Association. (2011). 'Letter to AFC regarding AFC U-16 Championship 2012 qualifier: protest on player's age eligibility.'

Soccerex. (2011a). 'Japan chief calls for AFC elections.' Retrieved from www.soccerex.com/industry-news/japan-chief-calls-for-afc-elections/ on 18 September 2011.

Soccerex. (2011b). '*K-League looks to the future.*' Retrieved from www.soccerex.com/industry-news/k-league-looks-to-the-future/ on 18 September 2011.

Soccerex. (2012). 'AFC bags sponsorship brace with Epson and Makita.' Retrieved from www.soccerex.com/industry-news/afc-bags-sponsorship-brace-with-epson-and-makita/ on 3 July 2012.

Somerford, B. and Kim, Y. (2011). 'K-League set for radical changes with promotion-relegation to be introduced as well as a "split system".' Retrieved from www.goal.com/en/news/14/asia/2011/10/05/2697324/k-league-set-for-radical-changes-with-promotion-relegation-to-be- on 11 December 2011.

Soosay, Alex (AFC General Secretary) interview by B. Weinberg, 28 September 2011, Kuala Lumpur.

Sugden, J. and Tomlinson, A. (1998). *FIFA and the contest for world football: Who rules the peoples' game?* Cambridge: Polity Press.

Sugden, J. P. and Tomlinson, A. (2003). *Badfellas: FIFA family at war.* Edinburgh: Mainstream.

Sugihara, Kaita (Acting Head of AFC Vision Asia Department) interviewed by B. Weinberg, 29 September 2011, Kuala Lumpur.

Tan, T.-C. and Bairner, A. (2010). 'Globalization and Chinese sport policy: The case of elite football in the People's Republic of China.' *The China Quarterly, 203*, 581–600.

Thaindian News. (2008). 'AFC spells out Chelsea's role in Vision India.' Retrieved from www.thaindian.com/newsportal/sports/afc-spells-out-chelseas-role-in-vision-india_10079749.html on 19 February 2011.

The Star. (2007). 'Red Devils' trip to Malaysia is off.' Retrieved from http://thestar.com.my/news/story.asp?file=/2007/5/9/nation/17672261&sec=nation on 18 September 2011.

The Straits Times. (2011). 'ASEAN Super League "in couple of years".' Retrieved from www.straitstimes.com/BreakingNews/Sport/Story/STIStory_683488.html on 16 July 2011.

Tokuaki, Suzuki (AFC Director of Competitions) interviewed by B. Weinberg, 27 September 2011, Kuala Lumpur.

Velappan, P. (2000a). 'Message from the General Secretary.' *Football Asia: The official magazine of the Asian Football Confederation*, September, p. 1.

Velappan, P. (2000b). 'Message from the General Secretary.' *Football Asia: The official magazine of the Asian Football Confederation*, November, p. 1.

Velappan, Peter (AFC General Secretary 1978–2007) interviewed by B. Weinberg, 11 October 2011, Kuala Lumpur.

Vietnam Football Federation. (2011). 'Letter to AFC regarding catalogue on potential infringements under the AFC Disciplinary Code.'

Warshaw, A. (2011a). 'It's official: Blatter v Bin Hammam for FIFA President.' Retrieved from www.insideworldfootball.com/news/1-latest-news/9009-its-official-blatter-v-bin-hammam-for-fifa-President on 5 April 2011.

Warshaw, A. (2011b). 'AFC hand out suspensions following Champions League semi-final brawl.' Retrieved from www.insideworldfootball.biz/worldfootball/asia/9846-afc-hand-out-suspensions-following-champions-league-semi-final-brawl on 11 December 2011.

WeAreAsia. (2012). 'The AFDP.' Retrieved from www.weareasia.com/en/the-afdp/ on 6 July 2012.

Weinberg, B. (2012). '"The future is Asia"? The role of the Asian Football Confederation in the governance and development of football in Asia.' *The International Journal of the History of Sport*, *29*(4), 535–552.

West Asian Football Federation. (no year). 'Organization.' Retrieved from www.thewaff.com/page.asp?pageID=2&parentid=&lang=eng on 21 June 2012.

Whittle, John (Head of AFC Grassroots and Youth Department) interviewed by B. Weinberg, 11 October 2011, Kuala Lumpur.

Wikipedia. (2012). 'South Asian Football Federation.' Retrieved from http://en.wikipedia.org/wiki/South_Asian_Football_Federation#cite_note-0 on 20 June 2012.

Williams, J. (2007). *A beautiful game: International perspectives on women's football*, Oxford: Berg.

World Sport Group (no year). 'Corporate fact sheet.' Retrieved from www.worldsportgroup.com/files/WSG_presskit.pdf on 1 October 2011.

Yusa, M. Y. (2011). 'Rebuilt academy in Myanmar opens.' Retrieved from www.the-afc.com/en/news-centre/inside-afc/793-dream-asia/37039-rebuilt-academy-in-myanmar-opens on 12 December 2011.

Zureikat, Fadi (General Secretary of the West Asian Football Federation) interviewed by B. Weinberg, December 2011, email interview.

6 AFC's framework and decision-making

The Congress

The Congress is defined as the supreme and legislative body of AFC. Being the forum in which all member associations meet and gather in order to cast their votes for important decisions, the Congress can be either an Ordinary or an Extraordinary Congress. While the Ordinary Congress takes place every two years, the Extraordinary Congress can be convened whenever the Executive Committee considers it necessary, or if the office of the President remains vacant for more than one year, or following a request by at least one-third of the member associations.

With the President being responsible for conducting the Congress as required by the Standing Orders of the Congress of AFC, all ordinary member associations have one vote and the possibility of making proposals, while having the right to be represented by a maximum of three delegates, one of whom shall cast the ballot on behalf of his or her association. It is mandatory that the delegates belong to only one member association and that they have been elected by the responsible body of the particular member association. Furthermore, it is prohibited to cast a vote by proxy or by letter. In addition to the delegates of the member associations, the Congress has the right to appoint observers and honorary officials; the former can participate but are not allowed to engage in debates or the voting process, while the latter may have the right to debate. Also, the Executive Committee can invite other persons to the Congress although they, like the other invitees, do not have the right to vote. Finally, the scope of the delegates is completed by the members of the Executive Committee and the General Secretary, all of whom do not have voting rights (Asian Football Confederation, 2011a: 21, 23, 25).

Regarding its authorities, the Congress covers a wide range of areas. First and most importantly it is responsible for adopting and amending the AFC Statutes, as well as the regulations governing the application of the Statutes and the Standing Orders of the Congress. It therefore has the authority to modify and amend the institutional core of AFC, namely its rules and regulations. Moreover, the Congress has authority over the election and revocation of the most important office bearers, namely the Executive Committee members, the FIFA Vice-President and the AFC President. As for the constituency of AFC, the Congress decides on whether to admit, suspend or even expel a member association. It is also in charge of approving the financial statements, the budget and the General Secretary's Activity Report. It has to fix the membership subscriptions and make decisions regarding awards and honorary titles. Finally, the Congress is the only body that may dissolve AFC (Asian Football Confederation, 2011a: 22).

With regard to the necessary quorum of the Congress, it is required that a simple majority of the member associations with the right to cast a vote are in attendance. In case this is not achieved, a second Congress shall be conducted within 24 hours. The Congress can generally only pass decisions if a simple majority of the valid votes is obtained. Votes are made either by showing hands, by a roll call or by acclamation, or in the case of a secret ballot through electronic devices. The election of office bearers, for instance, is to be made by secret ballots unless a candidate for a certain position is not opposed by anyone else. While the election of the President requires a two-thirds majority of votes in the first and a simple majority in the following ballot, the elections of the FIFA Vice-Presidents, the FIFA Executive Committee members, the AFC Vice-President and the AFC Executive Committee members require a simple majority in the first ballot. If no candidate is successful, further ballots take place with the candidate with the least number of votes in the prior ballot being eliminated (Asian Football Confederation, 2011a: 22–23, 59).

As for the procedure for the conduct of the Congress, it is important to note that, after the Executive Committee has fixed the date and place for the Congress, the member associations have to be informed not later than 90 days before the scheduled date. Not later than 30 days before the Congress takes place, the General Secretary is required to provide the member associations with the agenda, the report of the General Secretary, the financial statements and the auditor's report, the nominated candidates, proposals for amendments or alterations to the Statutes, as well as other proposals by the member associations or the Executive Committee. Proposals to amend or alter the Statutes have to be submitted in written form by either a member association (seconded by two others) or the Executive Committee. Amendments and alterations necessitate a three-quarters majority of the votes in order to be approved.

In principle all items on the agenda can be discussed. Being the chairman of the Congress, the President is in charge of permitting delegates to speak, closing discussions and also taking disciplinary measures against delegates, which, however, may be appealed against upon which the Congress has to make an immediate decision. According to the order, an item on the agenda shall be first stated by the chairman or an Executive Committee member, or by a representative of the respective committee or a representative of the member association that has asked for the item to be included on the agenda. Afterwards, the discussion is opened to other delegates, who can address the Congress in the order as they requested and only if permission has been granted. The decisions taken by the Congress take effect 30 days afterwards or on another date if this has been part of the decision (Asian Football Confederation, 2011a: 23–24, 26, 57–58).

As supreme and legislative body of AFC, the history of the Congresses includes important decisions and debates, bearing in mind the conflicts on issues of membership, especially during the 1970s. The following section, however, focuses on more recent examples, mainly drawing upon the official minutes of the respective Congress. In the last ten to 15 years, fundamental reforms have been undertaken with regard to changing and developing the Statutes, not least due to FIFA's objective of harmonising the institutional framework across the globe.

The 19th Congress took place in 2000 in Kuala Lumpur, where two important decisions were taken. While the Statutes were not altered, the Congress first approved the proposal by then Chairman of the Finance Committee, Mohamed Bin Hammam, to sell the marketing and television rights on a collective basis, and second elected Bin Hammam with 33 nominations and six seconds to serve as representative on the FIFA Executive Committee for the period from 2002 until 2004 (Asian Football Confederation, 2002a: 12).

At the Extraordinary Congress in 2001, no significant conflicts occurred but four important resolutions regarding cheating through fielding over-aged youth players, doping and drug abuse, corruption and bribery, as well as security matters in the stadiums, were unanimously approved by the delegates. Also, the Statutes were modified by including the necessity that executive bodies of member associations have to be elected properly, that the nominees for the AFC Executive Committee have to maintain a position in a member association, and that no members of the Executive or Standing Committees may serve in the Appeals and Disciplinary Committees (Asian Football Confederation, 2002a: 21–30, 36–63; 2002b: 6).

At the Extraordinary Congress in 2003, extensive changes were made to the Statutes, including new articles and renumbering of previously existing ones. Most notably it was decided that candidates for the post of President could be nominated by any two or more member associations, while the candidates were not allowed to originate from the same country as the retiring President (the latter, however, is not part of the current version of the Statutes). Also, it was adopted that each member association could only propose one candidate for each position in the Executive Committee, whose members (including the President) were henceforth to be under the age of 70 at the time of election. Moreover, changes were made with regard to the voting procedures and the composition of the Executive Committee, as well as the allocation of seats according to the different regions (Asian Football Confederation, 2003).

The 21st AFC Congress took place in 2004, when relevant decisions were taken with regard to the cycle of the Ordinary Congress and the tenure of the members of the Executive Committee. It was unanimously adopted that the 22nd Congress was to be held in 2007 and then ordinarily every two years. Accordingly, the term of office of the Executive Committee, which had originally been elected for the period from 2002 until 2006, was extended until 2007. The reason for doing so was to conform with FIFA's notion of not electing the President in the same year of the World Cup. For the same reason, the tenure of three AFC representatives on the FIFA Executive Committee was also extended until 2007, while the fourth, who was elected at the Congress in 2004, was to serve until 2009. These decisions implied for the next Ordinary Congress in 2007 that a total of ten Executive Committee members – the President, the FIFA Vice-President, two AFC Vice-Presidents, two representatives to the FIFA Executive Committee, the Honorary Treasurer and three other members – were to be elected for a four-year term, while another nine members – including two Vice-Presidents and seven other members – were to be elected for a two-year term. Thus, at the 23rd Congress the posts of the members expiring in 2009 – two Vice-Presidents, one representative on the FIFA Executive Committee and seven other members – would finally to be subject to election for a four-year period (Asian Football Confederation, 2004; 2007a: 9, 18).

The following year another Extraordinary Congress took place, in Marrakech, where amongst other changes to the Statutes an important amendment was unanimously approved, which provided the opportunity to revoke members of the Executive Committee if they did not take necessary responsibility for the post. Moreover, the Congress unanimously adopted the AFC Disciplinary Code and the Code of Ethics. The constitutions of the AFC Coaches Association and Referees Association were also adopted and their formations ratified. Finally, the Congress unanimously ratified the admission of the Australian Football Federation (AFF) as an ordinary member and approved a resolution to provide the Executive Committee with the mandate to decide on which region the AFF should belong to (Asian Football Confederation, 2007a: 22–23, 31).

In 2007 the 22nd AFC Congress took place in Kuala Lumpur. For the first time in the history of AFC the nominees for the Executive Committee were not contested and were unanimously elected by the delegates. Bin Hammam, who was also re-elected at this Congress, stressed in his welcome speech that the fact that the Executive Committee were to be elected unanimously had to be interpreted as a reflection of the unity and team spirit within AFC. With regard to the AFC Statutes, the Congress unanimously approved the following changes:

- the installation of the Internal Audit Committee with the aim of replacing the role and duties of the Honorary Treasurer with professionals (accordingly it was decided that the vacant seat on the Executive Committee would be made available for another member);
- the establishment of additional seats on the Executive Committee for four female members (one from each region as well as a post as Vice-President);
- the transfer of the Maldives into the South and Central region, since the AFF had been categorised in the ASEAN region;
- the transformation of the Ethics and Fairplay Committee, the Legal Committee, the Medical Committee and the Security Committee into Ad-hoc Committees or Bureaus whenever deemed necessary;
- the deletion of the Players' Status Committee because these matters would henceforth be covered by FIFA;
- and the introduction of a rotational basis between East and West Asia for the hosting of final competitions.

Furthermore, it was at this Congress that the delegates decided to pass a resolution according to which all member associations were not to invite any foreign team for friendly matches during the period of the Asian Cup taking place in July 2007. Despite this resolution, the FAM later scheduled an invitation for Manchester United which led to a conflict with AFC; based on the decision made by the Congress and an additional agreement that had been signed the FAM, however, saw itself under pressure and cancelled the match (Asian Football Confederation, 2009a: 7–8, 14–15, 18, 26–28; 2007b).

Three controversial aspects dominated the run-up to the 23rd Congress, which took place in May 2009. Following a meeting of the Executive Committee on 29 July 2008, the decision had been made to open a bid to all AFC member associations to host the headquarters of the confederation. For the purpose of defining the criteria for hosting the headquarters, the Executive Committee installed an ad-hoc committee which came up with requirements regarding the host's acceptance, economic considerations, the legal status, as well as financial support. Amongst others, these included specifically the recognition of AFC as an international organisation with diplomatic status for its senior officials, tax exemptions for all AFC revenues as well as employees, the provision of land without any cost for AFC, and an interest-free loan for the construction of the building. Given the long history of AFC being located in Malaysia, the decision to open a bid was not received positively by all member associations and also led to debates involving former officials such as Peter Velappan, as well as Malaysian politicians. Shortly before the Congress took place though, Bin Hammam met with the Malaysian Prime Minister in order to discuss the matter, which evidently resulted in the outcome that the President successfully asked the Congress to withdraw the item from the agenda.

The second controversial issue, meanwhile, arose from the question of whether the member associations of Afghanistan, Laos, Timor Leste and Kuwait would receive voting rights for the Congress. Afghanistan, Laos and Timor Leste had competed merely in the Festivals of Football, which were not regarded as actual competitions by the Executive Committee, and therefore it argued that these member associations did not qualify to vote. As for Kuwait, the situation was that the member association (the KFA) had been suspended due to not complying with the legal framework as required by FIFA and AFC, as a result of which the KFA as well as the government of Kuwait had engaged in resolving the matter by putting forward a roadmap for amending the national law and installing an appropriate committee structure. A couple of weeks before the AFC Congress took place, the KFA held elections which, however, the AFC Executive Committe interpreted as non-compliant with FIFA's instructions. FIFA, meanwhile, took the view that the KFA's election had fulfilled the required procedure and thus informed AFC accordingly on its recognition of the KFA's Statutes, the election's legality, and revocation of the suspension. Moreover, FIFA, as well as other Asian member associations and football officials, demanded AFC to provide the member associations of Afghanistan, Laos and Timor Leste with the right to vote in order to avoid further conflicts within the confederation, and to create a complete and fair democratic basis for the important items that were to be decided at the Congress. These items included – and this was the third and most important aspect – the election for a seat on the FIFA Executive Committee.

AFC's seat on the FIFA Executive Committee had belonged to Bin Hammam, but this year he was contested by Bahrain's Salman Bin Ebrahim Al Khalifa. The campaign period was characterised by disputes between the candidates, with Al Khalifa claiming that AFC had been split under the presidency of Bin Hammam and the latter threatening that if he lost the election he would step down as President. In addition, allegations and accusations of buying votes and abusing power circulated so that, in his speech to the Congress, FIFA President Blatter appealed to fair play and ordinary conduct. He also stressed that all football organisations and entities recognised by FIFA fall under Swiss law and therefore FIFA found it necessary to monitor the procedures through its General Secretary Jerome Valcke and a Swiss notary public. In this tense situation the Executive Committee opined to let the Congress decide on whether Afghanistan, Laos and Timor Leste should receive voting rights, which they finally did. As for Kuwait, the Executive Committee did not recommend the Congress to recognise the Executive Body of the KFA, but Bin Hammam took the initiative and proposed to the Congress to allow Kuwait the right to vote. Again, the Congress accepted and decided in favour of the member association. The election, meanwhile, ended with a close result: Bin Hammam received 23 and Al Khalifa 21 votes out of 46, with two votes having been invalid. Afterwards, Bin Hammam thanked the Congress for its support and its democratic way of functioning, while drawing attention to the fact that the election had had to be monitored by external persons and that therefore in future the member associations should work together to regain their trust in each other (Asian Football Confederation, 2008; 2010a: 13–15, 21, 37; 2009b; AFP, 2009; ESPN, 2009; Reuters, 2009).

A further important aspect on the agenda was the approval of the AFC's budget for the period from 2009 until 2012. Referring to the publication of the Financial Report 2007–2008 and the Budget Cycle 2009–2012, which had been disseminated to the member associations in advance of the Congress, Japan's Kohzo Tashima expressed his concern over approving the budget, arguing that this would include the approval of the planned agreement between AFC and the WSG covering a period until 2020. He therefore concluded that the Congress should not approve the budget before the delegates and members were provided with information on the planned contract. Richard Lai from Guam, who criticised the fact that in spite of higher income the expenditure on grassroots activities was not to be increased, shared this scepticism. He also expressed his concern about the expected negative budget of roughly US$4.5 million at the end of the year 2012, and he concluded by asking the Congress to reject the budget while also proposing that member associations should have more say with respect to how the budget should be apportioned. Following this statement, Skeikh Ahmad Falad Al Sabah from Kuwait turned to the Congress to say that the contract with the marketing partner should not run for a period of 12 years but instead for four, maybe being renewed afterwards. He thus proposed that AFC should distribute 20 to 30 per cent of the generated income equally amongst all member associations.

Responding to these aspects, President Bin Hammam referred to the AFC's payments of US$75,000 p.a. to 30 member associations as well as to the investments made through the Festivals of Football, Project Future and the competitions for which AFC covers the costs for boarding, lodging and the match officials. Furthermore, he stressed that the contract with the WSG is to last for eight years from 2013 until 2020 and that the current contract would end in 2012. Finally, he emphasised the importance of the AFC's two major competitions, the Champions League and the Asian Cup, as primary sources of income for the confederation. Afterwards, Manilal Fernando from Sri Lanka informed the Congress that the Executive Committee had in fact decided to approve the budget the day before, referred to the fact that the decision had been made unanimously, and argued that the contents of the contract with the WSG were an exclusive matter of the Executive Committee and not the Congress. Also, Ganesh Tapa from Nepal questioned the timing and motives behind the statements, saying that in the last six years neither members of the Congress nor the Executive Committee had expressed any concerns with regard to these aspects. This opinion was supported by Rahif Alameh, General Secretary of the Lebanese FA, who dismissed the questions as personal attacks on the President, urging the Congress to pay attention to the work ethics of the individuals rather than to their personalities. On the other hand, Abdul Aziz bin Abdul Rahman from Malaysia questioned if there were no possibilities of extending the negotiation talks regarding the marketing rights to other potential partners and therefore asked for cancelling the current procedure. Bin Hammam, however, put an end to the debate when stating that other options were impossible since AFC had already entered a commitment for eight years based on the decision of the Executive Committee. He also made clear that the contract was in the last phase of its completion and that, apart from that, the contract was not an item to be discussed according to the agenda of the Congress. Finally, the budget was to be decided upon and 22 delegates voted in favour while 14 others were against it (Asian Football Confederation, 2010a: 18–20).

Meanwhile, as for amendments to the AFC Statutes, the Korean Football Association proposed adding an article prescribing standards of good conduct for all AFC officials and employees. Ms. Dzung of the AFC's Legal Department was asked by Bin Hammam to comment and explain the position of the Executive Committee on the matter. She mentioned that the content of the proposal was already included in Articles 6 and 36 as proposed by the Executive Committee. Also, the Code of Ethics and the Disciplinary Code, as explained by Ms. Dzung, provided sufficient rules for cases of misconduct, infringements or violations. The KFA official Sam Ka, however, insisted on not withdrawing the proposal upon which FIFA in the person of Sepp Blatter was asked for its legal opinion, which was to withdraw the proposal to maintain consistency between FIFA's and AFC's Statutes and regulations. This intervention was followed by

Bin Hammam's proposal to discuss the Kuwait Football Association's request to establish a Legal Committee rather than continuing with decisions on further amendments. Accordingly, the Congress decided by acclamation to install a respective body which would then concern itself with and provide its view on the other proposals that had been submitted. Thus no extensive amendments were decided at the Congress (Asian Football Confederation, 2010a: 22; 2009c: 2–3).

The latest Congress for which minutes were available was the Extraordinary Congress in 2010, conducted in Johannesburg, South Africa, in the course of the World Cup tournament. After it had been decided in 2009 to postpone further amendments, this Congress could finally draw upon the work of the newly established Legal Committee. Accordingly, the AFC Vice-President and Chairman of the Legal Committee, Moya Dodd, outlined the major objectives the committee tried to achieve: first, to foster the alignment of AFCs and FIFA Statutes; second, 'to improve the clarity and certainty of the Statutes'; and third, 'to promote better governance and administration within AFC, for example the inclusion of the AFC President as a FIFA Executive Committee member'. For the purpose of preparing respective proposals for amendments, Ms. Dodd pointed out that the committee had worked together with various bodies such as the AFC Secretariat, its Legal Department as well as FIFA in order to ensure a high-quality output. She specifically drew attention to the correspondence with FIFA in two respects: first, FIFA criticised the AFC Statutes for being too restrictive in terms of requiring member associations to participate in at least three competitions within a period of two years. Ms. Dodd, however, remarked that the idea of this regulation was not to exclude but to encourage, and that the Congress may therefore decide in favour of the current statute. Second, she referred to FIFA's regulations according to which members of its Executive Committee were not eligible to act as delegates of their member associations with the right to vote at the Congress. She commented that the Legal Committee concluded it was not imperative to change the AFC's provisions accordingly but that FIFA preferred it. Finally, as for the other proposed amendments, Ms. Dodd said that the Legal as well as the Executive Committee had approved these and that it was now up to the Congress to decide. Afterwards Bin Hammam commented on the proposals, saying that he had consulted with the members of the Executive Committee with the result that it proposed to maintain the AFC's regulation according to which the Executive Committee members also hold the right to vote as delegates of a member association at the Congress. Moreover, Bin Hammam pointed out that it was his request that the article according to which the AFC President should automatically become a member of the FIFA Executive Committee should not be implemented with immediate effect but postponed to 2015, after the next presidential election in 2011. He then opened the floor for further comments which, however, merely addressed minor aspects such as the choice of words. Eventually the Congress decided with 44 votes in favour and no abstentions for the proposed amendments, including approval of FIFA's preference for the voting rights of the Executive Committee members as well as the request made by Bin Hammam. Overall, it can be said that this Congress can be considered very important due to the extensive amendments to bring alignment with FIFA's Statutes (incl. quote Asian Football Confederation, 2011b: 15–17, 19; 2010b: 20ff.).

Regarding the 2011 Congress, which took place in Doha in advance of the Asian Cup tournament, no minutes were available, merely the information provided through the AFC's website. Major items on the agenda were the election of the President, for which Bin Hammam ran unopposed, the position of FIFA Vice-President, competed for by Chung Mon-Joon and Prince Ali Bin Al Hussein, as well as two positions on the FIFA Executive Committee. As the only candidate, Bin Hammam was re-elected by the Congress delegates for another four years. Meanwhile, Prince Ali won the election for the post as FIFA Vice-President by 25 to 20 votes, a rather surprising outcome considering that his competitor Chung had been on the FIFA Executive Committee ever since 1994. In fact, Bin Hammam had backed Chung's candidacy and expressed his scepticism towards Prince Ali's involvement in the Olympic movement. Finally, due to the fact that Manilal Fernando from Sri Lanka and Worawi Makudi from Thailand were elected for the remaining seats on the FIFA Executive Committee, a significant power shift occurred with no East Asian member associations represented on the FIFA board. (This, however, changed in 2011 when Zhang Jilong replaced the suspended Bin Hammam as Acting President.) Bin Hammam later told the press that there had actually been initiatives to distribute the FIFA Executive Committee seats on a regional basis but the East Asian countries had blocked these (Asian Football Confederation, 2011c, 2011d, 2011e, 2011f; Corbett, 2011).

The Executive Committee

Defined as the executive body of AFC, the Executive Committee is comprised of 24 members including the President, one FIFA Vice-President, one female Vice-President, four Vice-Presidents, three additional representatives for the FIFA Executive Committee (including the AFC President unless he is a FIFA Vice-President), three female members and 12 other members. Whereas the President may come from any of the four zones that AFC is divided into, each of the four Vice-Presidents has to be from one of the four zones. For the members of the Executive Committee each zone has a quota, with West Asia, South and Central Asia as well as the ASEAN zone holding six seats, while East Asia has five. As for the representatives that are elected for the FIFA Executive Committee, they may come from any of the zones without having a limiting effect on the number of seats, though they are included in the quota.

The members of the Executive Committee are elected for a period of four years and they have the right to be re-elected. However, an Executive Committee member is not allowed to function as a member of the judicial bodies during the mandate. (See later in this chapter for a discussion about the judicial bodies.) Meanwhile, the AFC Congress has decided to reduce the maximum number of terms the President can be elected for to three (12 years). Also, all nominees for the Executive Committee including the President have to be under the age of 70 on the date of election. Of further importance is the article according to which a member association is not allowed to propose more than one candidate for each position on the board. In addition, a member association may have merely one representative in the committee. If a post becomes vacant, the President has the possibility of making a proposal to the Executive Committee upon which the members can appoint somebody until the next Congress takes place. As for the competences of the Executive Committee, the Statutes state that it is generally responsible for all matters which are neither covered by the Congress nor other bodies of AFC. This includes decisions on funding and financing in order to reach the goals the Executive Committee aims to achieve. Amongst its duties are the implementation of AFC's objectives, the preparation and convention of the Congress, ensuring the proper application of the Statutes, the approval of regulations regarding the internal organisation of AFC, the formulation of rules and regulations, the appointment of the Chairmen, Deputy Chairmen and members of the standing committees as well as the judicial bodies, the installation of ad-hoc committees, the approval and submission of relevant documents to the Congress (for example activity reports, financial statements, proposals for amendments or external auditing), the appointment and dismissal of the General Secretary as proposed by the President, as well as the provisional dismissal of a person or a body and the suspension of a member association until the Congress decides.

The Executive Committee is obliged to meet at least twice a year. Being responsible for convening the meetings, the President has to fulfil this duty within a period of 21 days if 50 per cent of the members ask for it. As for the agenda, each member has the right to suggest items and submit these to the general secretariat not later than six weeks in advance of the meeting so that the agenda can be forwarded to all members not later than four weeks before the date of the convention. While the meetings are not open to the public, the committee has the right to invite third parties, which may receive the permission to communicate their opinion on a particular matter. It is also noteworthy that a member can be provisionally suspended if he or she does not attend three consecutive meetings without being excused, or a total number of six meetings during the term; it is then up to the Congress to make a final decision regarding their suspension. However, members are permitted to attend a meeting via telephone or video conference provided the President and the General Secretary agree. In this context, the Executive Committee has the right to not only dismiss a person but also a body of AFC on a provisional basis until the next Congress. For this purpose any member may propose a dismissal as an item on the agenda of the Executive Committee.

Meanwhile, regarding the decision-making processes and respective requirements, the following aspects are relevant: the necessary quorum for an Executive Committee meeting is a simple majority of the elected members; the decisions require a simple majority of the votes given by the members participating in the particular meeting; in case of a tied vote the President has the deciding vote; the decisions have immediate effect unless the committee opines differently; and it is possible that actions may be decided upon in written form and without a meeting, but only if the majority of the members submit written approvals of this procedure.

A body with direct relevance in relation to the Executive Committee is the Emergency Committee. Its function is to replace the Executive Committee between two of its meetings in case a specific issue necessitates immediate decisions. It consists of the President and the five Vice-Presidents and, by analogy with the decision-making rules for the Executive Committee, its decisions have immediate effect but require ratification by the Executive Committee at its next meeting. Regarding the composition of the Emergency Committee, the President has the right to appoint a deputy for any member that cannot attend the meeting or has a conflict of interests; the deputy, however, has to be a member of the Executive Committee and come from the same zone (Asian Football Confederation, 2011a: 20, 26–31, 33).

From a historical perspective, it is interesting to take a look at the power dynamics within the Executive Committee before quotas for each zone were introduced. An example is the time of the bidding process for the Asian Cup in the mid-1990s. In the course of the tournament in the United Arab Emirates in 1996, the Executive Committee met in order to decide upon the bid for the next Asian Cup in the year 2000. Although not formalised, it was agreed that in principle the tournament should be awarded on a rotational basis. In 1996, however, the seven Arab members of the Executive Committee again backed a bid from the Middle East, namely Lebanon. Due to the fact that India and Sri Lanka supported the bid, South Korea and Hong Kong withdrew their candidatures and Malaysia merely presented theirs but asked not to be considered. Under these circumstances, the Executive Committee voted for Lebanon as host country with a majority of 14 to two, leading to enraged reactions from the Chinese. Also noteworthy is that Bin Hammam was elected to the Executive Committee with 17 votes to ten, while his rival Timothy Fok was essentially only backed by Hong Kong and China. Thus it could be argued that at that time the composition of the Executive Committee was rather beneficial for the Middle Eastern countries in terms of decision-making (Sugden and Tomlinson, 1998: 160).

In the first years of Bin Hammam's presidency, however, several structural and institutional reforms were brought on the way, including quotas for the zones when electing the members of the Executive Committee. Moreover, the structures within the administration were changed, which also had an effect on the decision-making procedures at the executive and committee levels. These transformations included the appointment of new officers, heads and directors, as well as a number of Assistant General Secretaries and a Deputy General Secretary; a selection of directors met with the President and the General Secretary on a weekly basis in order to discuss important matters and strategic proposals for the committee meetings. Generally, the committee meetings took place three to four times a year and the departments had to prepare an agenda for their respective committee's meeting, which was then sent to the President's office and forwarded to the committee members who in turn could add items of interest before the agenda was formalised by the General Secretary. In the meetings, the chairmen acted according to the agenda while Bin Hammam also attended the meetings and played a significant role. The items on the agenda were discussed and decided upon before being formalised for presentation to the Executive Committee on the following day. Each member of the Executive Committee received a folder with the relevant items discussed in a particular Standing Committee, and then there was in principle time for discussion and the decision-making process.

As pointed out by former AFC employees, the times during which Peter Velappan was General Secretary were characterised by extensive sessions with long discussions. Velappan was certainly a key actor in this context but he would have been open-minded enough to give adequate time to discussions and spontaneous statements, while Bin Hammam was very much focused on organising the committee meetings more efficiently in terms of concentrating strictly on the points on the agenda. Due to the fact that Bin Hammam was determined to professionalise the structures and processes within AFC, he conducted the Executive Committee meetings in a very goal-oriented manner, which was perceived as positive and necessary given the heterogeneous composition and membership of AFC and its bodies. On the other hand, his strict leadership style allowed him to influence certain processes and his strategic thinking enabled him to create situations in favour of his political ambitions; supposedly it was usually the case that the Executive Committee members were very supportive of his positions. Velappan stated that Bin Hammam acted 'like a dictator' and removed the freedom of speech, not least given the circumstance that the duration of the meetings of the Executive Committee was reduced from two days to 60 minutes without providing anybody with the chance to raise a point not mentioned on the agenda. He also criticised that decisions were basically a fait accompli made prior to the actual meetings through informal talks and exchanges in which Bin Hammam played an important part. Whether these accusations are accurate is difficult to verify; however, the fact remains that the structural conditions of the committee meetings were modified under Bin Hammam. With regard to the development of the power dynamics and competences of the Executive Committee, meanwhile, it can be said that the previously mentioned cases – such as the handling of the marketing deal with the World Sport Group, the idea to initiate a bidding process for hosting the headquarters of AFC and the definition of respective criteria, as well as the adoptions of internal regulations and codes – indicate not only the executive but even the legislative power of the Executive Committee (Al Sabah, 2011; Flath, 2012; Velappan, 2011).

As for looking at recent developments concerning the Executive Committee, it should be mentioned that, after Bin Hammam was suspended in 2011, Zhang Jilong in his role as Acting President was nominated as Chairman of the Executive Committee. It was also decided to provide Jilong with the seat on the FIFA Executive Committee previously held by Bin Hammam. In this context, the

Executive Committee approved the establishment of an ad-hoc committee dealing with the situation AFC was confronted with in order to fight corruption, improve the administrative structure and increase financial transparency. Finally, besides having reappointed Alex Soosay as General Secretary for the period from 2011 until 2015, other important decisions in 2011 and 2012 included the renaming of the AID27 Programme as the Financial Assistance Programme and the increase of the annual grant to US$250,000, as well as the reviewing of the headscarf rule with FIFA and International Football Association Board (IFAB) (Asian Football Confederation, 2011g, 2011h, 2011i, 2012).

The President

According to the AFC Statutes, the President is defined as the legal representative of the confederation. Amongst his duties are the following: the implementation via the general secretariat of decisions that have been taken by the Congress and the Executive Committee; securing the proper functioning of all AFC bodies; overseeing the activities of the general secretariat; managing the relations with the member associations, FIFA and other actors; proposing the appointment and dismissal of the General Secretary; participating in the Congress, the Executive Committee and other committee meetings; functioning as ex-officio member in all Standing Committees but not having the right to vote; and having an ordinary vote in the Executive and Emergency Committees, as well as a casting vote in case of a tied vote.

Regarding candidates for the presidency, it has to be noted that he/she is elected for a period of four years and can be re-elected two more times. A nominee has to be proposed by at least two member associations and the proposal has to be handed over to the general secretariat not later than 60 days in advance of the Congress. In turn, the General Secretary shall then inform the member associations about the respective proposals. Finally, as the President is the legal representative of AFC, he or she has the competence to sign on behalf of the organisation. The Executive Committee, meanwhile, has the right to set-up internal regulations with regard to a collective signature of officers, especially if the President is absent (Asian Football Confederation, 2011a: 31–32).

Bearing in mind the definition of the role of the President as mentioned in the AFC Statutes, it is important to shed light on how the post has been interpreted and performed by its bearers. Observers have stated that during the tenure of Peter Velappan as General Secretary, the Presidents played a rather representative role, given that in cases such as Sultan Ahmad Shah he was very much occupied by his role as ruler of the state. Velappan was therefore the person in charge of controlling AFC and he also gained a lot of power behind the scenes. However, when Bin Hammam became President in 2002 the patterns of leadership and the institutional framework changed with the consequence that, although Velappan had initially been one of his supporters, Bin Hammam began interpreting his role more as a hands-on President, in the course of which he minimised the competences of Velappan. This in turn laid the basis for the personal rivalry between the two football officials which culminated, for instance, in the heated debate over the idea of shifting the AFC headquarters and Bin Hammam's candidature for the FIFA Presidency.

In the context of the administrative reforms introduced under Bin Hammam, he sought to keep track of all decisions and policy issues. As a matter of fact he was always very well informed about most areas and had a genuine interest in implementing his vision of developing the game in Asia. His style of managing and promoting his ideas, however, created a source of conflict not only within the administration but also with regard to the relations with some of the member associations. Within the administration he established a structure in which almost all matters required his approval. He used to discuss all items with the group he had installed at least once a week, sometimes even twice, in one-and-a-half to three-and-a-half hour sessions according to a formal agenda. It also occurred that Bin Hammam dismissed people from the group or criticised them heavily if things did not work out as he wanted them to be, and he also undermined the authority of the directors so that certain processes were unnecessarily prolonged or even stagnated. On the other hand, the composition of the group was perceived as positive with regard to the fact that Bin Hammam knew about everything what was going on and expressed his interest in technical matters. Regardless, the management style affected the working atmosphere and the communication processes between the departments.

Meanwhile, in respect to the member associations, the presidency of Bin Hammam was defined by a highly politicised atmosphere in which he was accused of having ruled like an authoritarian leader. He was criticised for having divided AFC and having marginalised big member associations such as South Korea and Japan. Critics viewed his brainchild, the Vision Asia development programme, as an instrument used in order to gain support from a lot of the small member associations. The tensions became apparent not least in the context of the candidature for the seat on the FIFA Executive Committee in 2010, when Bin Hammam won in a very close election following a harsh and passionate campaign against his rival, Sheikh Salman Bin Ebrahim Al Khalifa.

When it came to Bin Hammam's announcement to run for the FIFA Presidency, he received wide support from a lot of Asian member associations. His critics, however, felt confirmed when Bin Hammam was suspended following accusations of bribery. Following the scandal the AFC executives, meanwhile, emphasised the need to improve transparency and accountability and to reduce corruption (Velappan, 2011; Careem, 2011; Flath, 2012; Gautam, 2011; Church, 2011; Hashim, 2011; AFP, 2008; Duerden, 2011; Warshaw, 2011; Dorsey, 2011; Bisson, 2011; *The Star*, 2011).

The standing committees and ad-hoc committees

The primary function of the standing and ad-hoc committees is to support the Executive Committee in the delivery of its responsibilities and obligations, as formulated in the Statutes and regulations defined by the Executive Committee.

According to the Statutes recognised by the Congress in 2011, AFC incorporates standing committees for the following areas: finance and marketing, internal audit, competitions, technical matters and Vision Asia, coaching, referees, women's football, Futsal, social responsibility, legal issues, and medical aspects. Each committee consists of a chairman, a deputy chairman, and at least three and at most nine other members. It is required that the chairmen are members of the Executive Committee, apart from the Chairman of the Internal Audit Committee. Following proposals by either the member associations or the President, all members are appointed by the Executive Committee. While the tenure lasts four years, the members can be reappointed but also dismissed if a member does not attend three consecutive or a total of five meetings during the term without being excused (in the latter case it is up to the Executive Committee to make a final decision).

Regarding the dates and agenda setting, the general secretariat is responsible for this procedure and consults with the chairman in question on the specific items. As for decisions taken or formulated by the standing committees, they have to accord with the viewpoint of a delegation from the Executive Committee or have to be ratified afterwards by the Executive Committee so that they become final. The necessary quorum for the Standing Committees has to be at least 50 per cent of the number of members that have been appointed, while it is possible – by analogy with the regulations for the Executive Committee meetings – to take decisions without an actual meeting but instead in written form. Furthermore, members are not permitted to cast their vote if the matter refers to the member association that the member represents or in any other case of conflict of interests. Of further importance is the option for each committee to install a bureau or a sub-committee drawing upon its members in order to resolve immediate or particular matters. Finally, each committee has the right to submit proposals to the Executive Committee with regard to amending or modifying the internal regulations which govern the procedures for the committees (Asian Football Confederation, 2011a: 33–35).

In regard to the specific competences of the standing committees, the Finance and Marketing Committee shall be examined first. Its primary competence is the monitoring of the financial administration of AFC, as well as an advisory service to the Executive Committee on all matters relating to finance. In addition, it is responsible for analysing the budget and the financial statements as proposed by the General Secretary for the Executive Committee and the Congress. Finally, it advises the Executive Committee on matters relating to marketing strategies and policies including negotiations and the implementation of contracts.

The second committee, namely the Internal Audit Committee, is concerned with the accuracy of the financial accounting, reviewing the report of the external auditor in case the Executive Committee asks for it, as well as conducting analyses and giving advice regarding the internal control systems and risk management policies.

The Competitions Committee, meanwhile, is responsible for the organisation of all AFC competitions and takes the decisions necessary with regard to all related aspects. The Technical and Vision Asia Committee is primarily concerned with expanding and improving the educational activities of AFC as well as overseeing the implementation of the Vision Asia development programme. It is also responsible for the AFC's Financial Assistance Programme, formerly known as AID27.

The Coaching Committee, meanwhile, deals specifically with the development of coach education schemes and programmes, in the context of which it promotes exchange programmes for both coaches and instructors between different member associations, as well as the improvement of the level of coaching in all countries and the development of instructors at all levels.

The Referees Committee's competence lies in the implementation and interpretation of the Laws of the Game; it can make proposals for amendments to the Executive Committee, which in turn can recommend these to FIFA. Furthermore it is responsible for appointing referees, their assistants and fourth officials, assessors and instructors for AFC matches. Finally, it approves annually the panel of elite referees, assistants and instructors.

While the Women's Committee deals with all relevant aspects of the women's game, the Futsal Committee is concerned with all Futsal-related matters. The so-called Social Responsibility Committee, meanwhile, is dedicated to all activities regarding social responsibility within AFC and the Asian football community.

The Legal Committee's primary function is to inform and advise the bodies of AFC on all legal matters relating to aspects of the game and the administration, Statutes and regulations of AFC and its member associations; moreover, it informs the Executive Committee on important issues that have to be dealt with by the Executive Committee, the Congress or other bodies, as well as providing assistance to the Executive Committee, the President and the General Secretary for assessing advice or services AFC has received from its legal advisors.

Finally, the last standing committee mentioned in the Statutes is the Medical Committee, which is responsible for all medical matters related to football, implying anti-doping.

As has been mentioned before, however, the Executive Committee has the option of installing ad-hoc committees for the purpose of dealing with specific issues within a limited period of time. In this respect the Executive Committee has the competence to decide on the composition of a particular ad-hoc committee and its tenure (Asian Football Confederation, 2011a: 35–37).

The judicial bodies

AFC has the right to install three judicial bodies: the Disciplinary Committee, the Appeals Committee and the Ethics Committee. As has been pointed out previously, the members of these bodies are not allowed to serve on any other body during their tenure. Both the Disciplinary and the Appeals Committee are composed of a chairman, a deputy chairman and a number of additional members the Executive Committee can decide upon; it is required that the chairman and the deputy chairman hold legal qualifications.

While the primary function of the Disciplinary Committee is to sanction and discipline member associations, officials, clubs and players as required by the Statutes, regulations and codes under consideration of the disciplinary competences held by the Congress and the Executive Committee, the Appeals Committee is designed to deal with appeals against decisions taken by the Disciplinary Committee, provided these have not been declared final in accordance with the Disciplinary Code. Regarding the range of sanctions that can be imposed in case of violations, AFC differentiates between natural and legal persons. Amongst the different forms of sanctions are warnings, fines, confiscations, expulsions, suspensions, transfer bans, annulments of results, and deductions of points or relegations.

The AFC Statutes clearly state that member associations, clubs, officials and players shall handle disputes within the confines of the legal bodies of the sport system rather than turning to public courts. As a matter of fact, AFC requires that member associations and registered clubs incorporate respective regulations in order to prevent officials and players from consulting ordinary courts. The highest legal body in the sport system recognised by AFC is the Court of Arbitration for Sport (CAS). In this sense the CAS Code of Sports-related Arbitration serves as a framework for the proceedings, while the AFC Statutes and regulations as well as Malaysian law shall be taken into consideration in case these become relevant. The CAS, however, may only be consulted when all other internal procedures have been made use of. Accordingly the decisions taken by the CAS have then to be considered as final and authoritative. Furthermore, it is important to note that the CAS can function as both ordinary court of arbitration and an appeals arbitration body. With regard to the latter function, the CAS may, however, only be consulted by directly affected parties unless the decision concerns doping-related matters, in the case of which WADA and FIFA may appeal to the CAS (Asian Football Confederation, 2011a: 38–42; 2010: 36–40; 2011j: 4).

The general secretariat and the General Secretary

Codified as the administrative body of AFC, the general secretariat is concerned with all administrative and operational matters within AFC. Located at the headquarters in Kuala Lumpur, it consists of various departments which mostly correspond with the thematic areas covered by the standing committees. The general secretariat is run by the General Secretary as defined by internal regulations regarding the organisational structures and procedures within AFC. As Chief Executive of the general secretariat, the General Secretary operates under the Executive Committee's direction. Amongst his duties are the following:

- the implementation of decisions taken by the Congress and the Executive Committee in accordance with the directives given by the President;
- the administration of the general secretariat;
- the appointment of staff and proposing managerial staff to the President;

- assisting and participating in the Congress and meetings of the Executive, as well as other committees;
- the responsibility for taking the minutes of all meetings, taking care of the AFC's publications and all correspondence;
- communicating and working together with FIFA, the confederations and all other actors in the context of promoting and achieving the goals of AFC;
- and managing the AFC's accounts as well as signing decisions on behalf of the committees, unless internal regulations prescribe differently.

As an ex-officio official, the General Secretary has to attend all meetings but does not have the right to cast a vote. Also, the General Secretary cannot function as delegate at the Congress nor can he or she be a member of any of the AFC's bodies (Asian Football Confederation, 2011a: 37–38).

As has been mentioned before, the administrative body of AFC has undergone rapid transformations in terms of quantity and quality. Especially during the presidency of Bin Hammam, ongoing restructuring processes have been implemented with regard to departments, personnel, formal procedures, internal regulations, and specific competences and responsibilities. For the purpose of this chapter, the AFC's Activity Report for the years 2009 and 2010, as well as several interviews with directors and heads of departments, shall serve as the basis for briefly portraying the departments of AFC.

The first department to be outlined is the Competitions Department. It constitutes the centre of all activities and developments regarding the AFC's competitions. These include the drafting of rules and regulations, holding the draws, registering players and officials, and preparing and implementing the competitions. Additionally, the department organises courses for match commissioners and club representatives. It is also in charge of the AFC Professional Football Project, in the course of which special mission teams undertake inspections in countries that may meet the criteria for participating in the AFC Champions League. One of the major tasks of the department is to develop a competitions calendar for the member associations participating in the Asian Cup, the ACL, the Challenge Cup, the President's Cup, the youth tournaments, the Futsal competitions and the qualifying rounds (Tokuaki, 2011; Asian Football Confederation, 2011k: 14).

The next department is the Women's Department, which is defined as a 'mini-AFC' in relation to all matters in the context of the development of the women's game in Asia. The responsibilities include not only educational schemes but also the preparation and organisation of the competitions, including the Women's Asian Cup, the youth tournaments, the qualifying rounds as well as the Girls Festivals for the Under-13 and Under-14 age groups. All these activities are managed in consultation with the Competitions Department (Teo, 2011; Asian Football Confederation, 2011k: 14).

Meanwhile, the Futsal Department is responsible for all Futsal-related policies and activities within AFC in terms of competitions, education and development programmes. It also engages in providing the member associations with guidelines and support in order to foster the standards and development of Futsal in the countries (Targholizade, 2011; Asian Football Confederation, 2011k: 14).

In the technical area, the so-called Education Division deals with educational matters with regard to coaches, referees and officials alike. Through the organisation of seminars, workshops and courses, it aims at providing all actors with the necessary knowledge on recent technical and administrative developments. The qualification of as many instructors as possible follows the objective to produce important know-how transfer and a multiplier effect (Al Sabah, 2011; Asian Football Confederation, 2011k: 15).

The Vision Asia Department, meanwhile, is specifically in charge of realising the AFC's football development programme in cooperation with the particular member associations. The measures include areas such as the administration of the member associations and clubs, the installation and establishment of clubs and leagues, grassroots and youth football, coach education and refereeing. For this purpose the department works together with the Education Division and the Referees and Grassroots and Youth Departments in order to conduct assessments and prepare concepts, guidelines and management strategies for the implementation of the projects (Sugihara, 2011; Asian Football Confederation, 2011k: 15).

The Grassroots and Youth Department, on the other hand, sets its focus on children and youth players and coaches with the aim to enhance their fundamental skills and techniques, as well as mindsets, as a basis for their career path. A key element of the department's work are the so-called Festivals of Football which take place in all AFC zones and are designed for the Under-13 and Under-14 age levels, as well as educating young coaches and referees (Whittle, 2011; Asian Football Confederation, 2011k: 15).

The Referees Department is primarily responsible for taking care that matches are conducted in accordance with the Laws of the Game. Therefore the department deals with appointing the referees for all AFC competitions, as well as defining and monitoring the standards of refereeing across the continent. Furthermore, it is in charge of implementing educational schemes for Asian referees and enhancing their development in collaboration with the member associations and other stakeholders. An example of such initiatives is the Elite Education Seminar for all elite officials. The general teaching materials used for the courses and seminars are produced in six languages and include videos, interactive elements and interpretations of the rules and regulations (Kashihara, 2011; Asian Football Confederation, 2011k: 15).

The third thematic area covered by the AFC's administrative bodies encompasses media and IT-related matters. The Website Department, for instance, takes care of the AFC's internet appearance. Produced in seven languages, the website contains the most recent news and information with regard to the AFC's activities, competitions and events. The Administration System and the Information Services Departments, meanwhile, are dedicated to maintaining the AFC's database of players, officials and competitions. Specifically, this implies securing a reliable flow of information for all departments and bodies working with the system.

The AFC Audio-Visual Unit, in the meantime, is in charge of managing the recording of all AFC matches, as well as events including the Congress or the Education Seminars. The purpose of this activity is not only to archive and systematise the material but also to process it for technical analyses, referees' assessments and coach education purposes (Asian Football Confederation, 2011k: 14–15).

The fourth sector is devoted to financial and marketing matters, which are covered by the Finance, Internal Audit and Marketing Departments. While the former deals with keeping record, managing the finances and planning the budget, the Internal Audit Department is responsible for the internal control system of AFC. Reporting to the Internal Audit and Executive Committees in order to guarantee objective conduct, the department provides recommendations for optimisation and advancement. The Marketing Department focuses on enhancing the commercial value of the AFC's products and competitions. In order to boost the image of AFC and its member associations, the department seeks to improve the relations with marketing partners and sponsors (Kuan, 2011; Saito, 2011; Asian Football Confederation, 2011k: 14–15).

A further important function, meanwhile, is performed by the Event and Logistics Department, which is concerned with all logistical aspects of the AFC's events and tournaments. It basically attends to all transport, travel, protocol and accommodation-related matters that become relevant in the context of the organisation of all AFC competitions, the Congress and committee meetings, as well as the Annual Awards. In the meantime, the Human Resources Department is responsible for taking care of all employment-related issues, including recruitment procedures, training of personnel and compliance with Malaysian law where applicable in order to ensure an effective employment structure within the organisation (Asian Football Confederation, 2011k: 15).

Last but not least, the Legal Department of AFC constitutes the administrative organ that is concerned with all legal issues. Functioning as the administrative branch of the Legal Committee, it assists with reviewing and preparing changes to the institutional and legal framework of AFC. For instance, it played an important role in the preparation of the extensive modifications of the new Statutes which were passed by the Congress in 2010, as well as the AFC Disciplinary Code and the Anti-Doping Regulations. It also provides support to the judicial bodies with regard to disciplinary and appeal cases. Moreover, it collaborates with FIFA's Legal Department in order to oversee the Statutes of the member associations and their accordance with FIFA's and AFC's requirements (Johnson, 2011; Asian Football Confederation, 2011k: 15).

Interim conclusions

The procedures and decision-making processes, as well as causal mechanisms of deliberation and interaction, require assessment of the AFC's institutional framework and the organs, bodies and individual representatives.

The most important decisions are taken at the AFC Congress, which is described as the supreme and legislative body. Here, all member associations convene in order to decide on the Statutes (requiring a three-quarters majority), membership affiliations and to elect representatives. All member associations have one vote and have the right to make proposals for the agenda of the Congress. Also, the Executive Committee may make proposals at the Congress. In the last 10 to 15 years, major reforms and amendments have been made to the Statutes against the backdrop of FIFA's objective to harmonise the institutional framework. This includes the establishment of a disciplinary catalogue and the Code of Ethics. Also noticeable is the fact that the Congress has been a site for political power games and for signalling the ambitions of individuals. Interestingly, however, the formal procedures are usually carried out without conflict and decisions are often unanimously adopted. The informal agenda setting and exertion of influence therefore take place behind the scenes.

The Executive Committee is the executive body of AFC. Its composition is based on a quota for each zone. Each member association may only have one representative in the committee. The committee is responsible for all matters not covered by the Congress or other bodies. This includes decisions on funding and finance matters, and the provisional dismissal of a member association, person or body. Each committee member has the right to propose an item for the agenda, decisions are made through a simple majority and the President has the deciding vote in case of a tied vote. The committee has the possibility of installing an Emergency Committee if decisions have to be taken urgently. Noticeable is that the committee plays de facto a strong role in influencing the legislative outputs. Throughout his presidency, Bin Hammam professionalised the procedures and made them more effective. However, it was also reported that he played a very strong role in the decision-making processes.

The President, who is defined as the legal representative of AFC, can have a crucial stake in preparing and influencing the decisions of the Congress and the Executive Committee. Besides having the right to make proposals for appointing the General Secretary and the chairmen of the standing committees, he is an ex-officio member of the Executive Committee and the standing committees. He has the deciding vote in the Executive Committee and can interpret his role as a leading figure with regard to discussing items on the agenda. Moreover, he can make an impact on the working procedures in the general secretariat of AFC.

The standing committees and the ad-hoc committees have the function of supporting the Executive Committee in the delivery of its tasks. Therefore, each standing committee is responsible for a specific thematic area. Their members are appointed by the Executive Committee and the chairmen have to be members of the Executive Committee. The decisions taken by the standing committees necessitate ratification by the Executive Committee.

The judicial bodies of AFC encompass the Disciplinary Committee and the Appeals Committee. Moreover, AFC has the option of installing an Ethics Committee. These organs deal with disciplinary matters including sanctions and appeals against decisions that have been taken. The committees operate within the confines of the sport law infrastructure, with CAS being the highest body. It is noteworthy that the decisions taken by the AFC's judicial bodies have no immediate and hardly any indirect impact on the institutional framework of the confederation.

Finally, the general secretariat is the administrative body of AFC. Its various departments relate to the areas the standing committees are in charge of. In this sense, the general secretariat is primarily responsible for the day-to-day operations of the confederation. The head of the administrative body is the General Secretary, who is under the direction of the Executive Committee. The general secretariat has undergone a rapid expansion throughout the last two decades, a development reflecting the growth of AFC, its range of tasks and outreach.

References

AFP. (2008). 'Hammam lashes out at Velappan.' Retrieved from http://thestar.com.my/sports/story.asp?file=/2008/11/27/sports/2654518&sec=sports on 19 February 2011.

AFP. (2009). 'Football: Velappan warns AFC members may quit.' Retrieved from www.channelnewsasia.com/stories/afp_sports/view/427143/1/.html on 18 September 2011.

Al Sabah, Hasan (AFC Deputy General Secretary and Director of Education) interviewed by B. Weinberg, 7 October 2011, Kuala Lumpur.

Asian Football Confederation. (2002a). *AFC XXth Congress 2002: Agenda*.

Asian Football Confederation. (2002b). *AFC XXth Congress 2002: Reports and activities (2000–2002)*.

Asian Football Confederation. (2003). *Amendments to the AFC Statutes adopted by the AFC Extraordinary Congress in Doha 2003*.

Asian Football Confederation. (2004). *AFC XXIst Congress 2004: AFC Executive Committee proposal to the 21st AFC Congress 2004*.

Asian Football Confederation. (2007a). *AFC XXIInd Congress 2007: AFC XXIst Congress 2004 minutes and AFC Extraordinary Congress 2005 minutes*.

Asian Football Confederation. (2007b). *AFC XXIInd Congress 2007: Nominations*.

Asian Football Confederation. (2008). 'Criteria for AFC headquarters.' Retrieved from www.the-afc.com/uploads/Documents/common/cms/afc/criteriahq.pdf on 5 October 2011.

Asian Football Confederation. (2009a). *AFC XXIInd Congress 2007: Agenda and minutes*.

Asian Football Confederation. (2009b). *Correspondence between FIFA, AFC and Kuwait FA on suspension of KFA*.

Asian Football Confederation. (2009c). *AFC XXXIIIrd Congress 2009: Proposals and secondments for Congress agenda*.

Asian Football Confederation. (2010a). *AFC Extraordinary Congress 2010: Agenda 2010 and minutes of AFC 23rd Congress 2009*.

Asian Football Confederation. (2010b). *AFC Extraordinary Congress 2010: Proposed amendments to the Statutes*.

Asian Football Confederation. (2010). *AFC Disciplinary Code*.

Asian Football Confederation. (2011a). '*Statutes of the Asian Football Confederation: Regulations governing the application of the Statutes and standing orders of the Congress*', 2011 edition. Retrieved from http://image.afcpresident.com/upload/library/AFC_Statutes_2011.pdf on 10 November 2010.

Asian Football Confederation. (2011b). *AFC XXIVth Congress 2011: Agenda 2011 and minutes of the Extraordinary Congress 2010.*

Asian Football Confederation. (2011c). 'All set for the Congress.' Retrieved from www.the-afc.com/en/afc-congress-news/32011-all-set-for-the-congress- on 30 October 2012.

Asian Football Confederation. (2011d). 'Bin Hammam wins third term as AFC President.' Retrieved from www.the-afc.com/en/afc-congress-news/32042-bin-hammam-wins-third-term-as-afc-President on 30 October 2012.

Asian Football Confederation. (2011e). 'Prince Ali wins FIFA Vice-President vote.' Retrieved from www.the-afc.com/en/afc-congress-news/32044-prince-ali-wins-fifa-vice-president-vote on 30 October 2012.

Asian Football Confederation. (2011f). 'Fernando, Worawi win FIFA Exco seats.' Retrieved from www.the-afc.com/en/afc-congress-news/32045-fernando-worawi-retain-fifa-exco-seats on 30 October 2012.

Asian Football Confederation. (2011g). 'Decisions of the AFC Executive Committee.' Retrieved from www.the-afc.com/en/media-releases/35884-decisions-of-the-afc-executive-committee on 18 September 2011.

Asian Football Confederation. (2011h). 'Message of unity from AFC ExCo.' Retrieved from www.the-afc.com/en/news/35880-message-of-unity-from-afc-exco on 18 September 2011.

Asian Football Confederation. (2011i). 'AFC to put best foot forward.' Retrieved from www.the-afc.com/en/news/37255-afc-to-put-best-foot-forward on 12 December 2011.

Asian Football Confederation. (2011j). *AFC Code of Ethics.*

Asian Football Confederation. (2011k). *AFC XXIVth Congress 2011: Activity Report 2009–2010.*

Asian Football Confederation. (2012). 'AFC ExCo takes important decisions.' Retrieved from www.the-afc.com/en/news/38461-afc-exco-takes-important-decisions on 27 June 2012.

Bisson, M. (2011). 'Monday briefing: Zhang wants Asian football's top kob; Man City ink YouTube deal.' Retrieved from www.worldfootballinsider.com/Story.aspx?id=34724 on 11 December 2011.

Careem, Nazvi (sports journalist and AFC Media Director 2008–2011) interviewed by B. Weinberg, 1 October 2011, Kuala Lumpur.

Church, Michael (sport journalist and Editor of *AFC Magazine* 1995–2001) interviewed by B. Weinberg, 26 September 2011, telephone interview.

Corbett, J. (2011). 'Asian football chief Bin Hammam rues Chung's FIFA departure.' Retrieved from www.afcpresident.com/en/News/view_news.php?pageNum=2&menuNum=24&idx=153 on 19 September 2011.

Dorsey, J. M. (2011). 'Asian Football Federation moves to dismantle Bin Hammam's legacy.' Retrieved from http://mideastsoccer.blogspot.com/2011/08/asian-football-federation-moves-to.html on 17 September 2011.

Duerden, J. (2011). 'Bin Hammam demands united Asia as Korea joins up behind Presidential bid.' Retrieved from www.worldfootballinsider.com/Story.aspx?id=34252 on 5 April 2011.

ESPN. (2009). 'Hammam survives close election.' Retrieved from www.espnstar.com/football/other-football/news/detail/item262670/Hammam-survives-close-election/ on 18 September 2011.

Flath, Thomas (AFC Director of Grassroots and Youth 2003–2006) interviewed by B. Weinberg, 13 February 2012, Mönchengladbach.

Gautam, Biplav (AFC Vision Asia Officer 2007–2009) interviewed by B. Weinberg, 14 September 2011, telephone interview.

Hashim, Rizal (Malaysian sports journalist) interviewed by B. Weinberg, 27 September 2011, Kuala Lumpur.

Johnson, James (AFC Director of International Relations and Development) interviewed by B.Weinberg, 11 October 2011, Kuala Lumpur.

Kashihara, Joji (Head of AFC Referees Department) interviewed by B.Weinberg, 23 September 2011, Kuala Lumpur.

Kuan, Bryan (Head of AFC Finance Department) interviewed by B. Weinberg, 30 September 2011, Kuala Lumpur.

Reuters. (2009). 'AFC election: Bitter Asian power battle reaches pivotal climax.' Retrieved from http://soccernet.espn.go.com/news/story?id=643454&sec=global&cc=5739 on 10 November 2010.

Saito, Satoshi (Head of AFC Marketing Department) interviewed by B. Weinberg, 10 October 2011, Kuala Lumpur.

Sugden, J., and Tomlinson, A. (1998). *FIFA and the contest for world football: Who rules the peoples' game?* Cambridge: Polity Press.

Sugihara, Kaita (Acting Head of AFC Vision Asia Department) interviewed by B. Weinberg, 29 September 2011, Kuala Lumpur.

Targholizade, Ali (AFC Director of Futsal) interviewed by B. Weinberg, 30 September 2011, Kuala Lumpur.

Teo, Julie (Head of AFC Women's Department) interview by B. Weinberg, December 2011, email interview.

The Star. (2011). 'FIFA VP Prince Ali calls for AFC election.' Retrieved from http://football.thestar.com.my/story.asp?file=/2011/7/27/football_latest/20110727101727&sec=football_latest on 18 September 2011.

Tokuaki, Suzuki (AFC Director of Competitions) interviewed by B. Weinberg, 27 September 2011, Kuala Lumpur.

Velappan, Peter (AFC General Secretary 1978–2007) interviewed by B. Weinberg, 11 October 2011, Kuala Lumpur.

Warshaw, A. (2011). 'Bin Hammam's Presidential hopes boosted by Asian support.' Retrieved from www.insideworldfootball.com/worldfootball/42-news/9071-bin-hammams-presidential-hopes-boosted-by-asian-support on 6 May 2011.

Whittle, John (Head of AFC Grassroots and Youth Department) interviewed by B. Weinberg, 11 October 2011, Kuala Lumpur.

7 AFC's policies, strategies and activities

Jurisdiction and regulation

According to its own description, AFC's core competence lies in 'running football in Asia', which is substantially connected with regulatory matters as well as passing new laws and implementing and enforcing these, primarily in the context of the AFC's competitions but also with regard to developing grassroots, youth and professional football structures across Asia (Asian Football Confederation, 2011).

Recognised by FIFA as one of its six confederations AFC is incorporated into the global infrastructure of football as defined by the world governing body, albeit AFC has possibilities to participate in shaping the institutional framework through its representatives and members at the global level. Consequently, AFC and its members are obliged to comply with FIFA's regulations including first and foremost the Laws of the Game, as well as Statutes regarding matters of administration and 'governance' (FIFA, 2011: 17–19). AFC would therefore work together with FIFA, member associations and governmental actors in order to enhance compliance. Due to the dual membership of national associations in FIFA and AFC, both consider it important to follow mutual approaches and objectives. The usual process is that AFC's legal personnel consult with their FIFA counterparts, after which a mission team is sent to the country in question to assess the situation. Given that Asia is a large and heterogeneous continent in political and legal terms there are different legal systems and national laws which, from the perspective of FIFA and AFC, is perceived as challenging inasmuch as both are determined to create a unified system, for instance with regard to player transfers, and for example in Thailand where the national law prevents the President of the member association from holding the post for longer than two years while FIFA requires a period of four. The talks with Presidents, Executive Committee and Congress members of national associations and governmental representatives thus require a high degree of diplomatic sensitivity and sophisticated negotiating skills. Regarding cases of government interference, AFC strives to negotiate with all actors involved and then make recommendations to FIFA on how to handle the conflict. In 2011, for instance, AFC contacted the Bahraini government to enquire into the allegations of players having been detained for political reasons, instead of directly taking measures without assessing the situation in advance. Generally, AFC is interested in finding a solution because if a member association is suspended by FIFA it is automatically also suspended from AFC. Hence AFC seeks to prevent such scenarios from occurring in the first place. It is noteworthy that while FIFA and AFC try to improve and maintain a unified system there maybe some exceptions such as China, where the CFA does, however, work according to the regulations (Johnson, 2011).

The AFC Statutes constitute the legal centrepiece of the organisation, covering all essential aspects in relation to its objectives, purpose, legal status, members, bodies and decision-making procedures, financial matters, competitions and club licensing. The Statutes have experienced various modifications and extensions in the last few years with the overall aim of aligning AFC with FIFA's requirements. While there may be some deviations, the idea is that fundamental points, especially with regard to organisational matters, are complied with. Differences exist for instance in respect to the standing committees due to the fact that each confederation has different thematic focal points. The statutory developments have led to organisational restructuring in the form of new standing committees, including the inception of the Legal Committee which replaced an ad-hoc committee. AFC has also been looking into creating a Member Associations Committee in order to generate more expertise on respective matters, rather than always taking these to the Executive Committee where delegates may not necessarily be as knowledgeable (Asian Football Confederation, 2011b; Johnson, 2011).

Furthermore, the AFC's authority grounds on two important legislative texts, the AFC Disciplinary Code and the AFC Code of Ethics. The basic functions of the Disciplinary Code are the definition of infringements of rules set-up and adopted by AFC in its regulations, in particular the Laws of the Game as prescribed by FIFA, and the determination of respective sanctions and disciplinary matters. Furthermore, it defines the organisational and regulatory function of the judicial bodies as well as the procedures that have to be accorded with. With regard to the scope of application, the Disciplinary Code covers all matches and competitions organised by AFC as well as breaches of the AFC Statutes relating to forgery, corruption and doping abuse. Subject to the code are natural and legal persons including member associations, clubs, officials, players, match officials, agents and spectators. The scope of possible sanctions, meanwhile, includes amongst others warnings, reprimands, fines, confiscations, cautions, expulsions, match suspensions, bans on engaging in any football-related activities, transfer bans, playing matches without spectators, result annulments, exclusions from competitions and deductions of points (Asian Football Confederation, 2010a: 11, 14–15).

Another important legal framework is provided through the AFC Code of Ethics. It has been codified with the distinct aim to protect the image of AFC and its objectives against unethical behaviour by officials. In this sense it is thought of as a code providing a basis for maintaining integrity and transparency. According to the definition, officials include representatives, members of committees, managers, coaches, match officials, medical officials, general staff as well as other personnel who hold responsibility for technical, medical or administrative matters within AFC itself, its member associations, leagues or clubs. In some cases the code can be applicable to players as well. The basic idea of the code grounds on the expectation that officials shall perform their duties with an ethical attitude, in accordance with the principles and objectives of AFC. Hence they are not allowed to abuse their power and position, particularly in order to create personal benefit. In the event of non-compliance with the code, an official may be sanctioned by the Disciplinary Committee, which can result in being dismissed from the position in question. The scope of potential infringements includes cases of conflict of interest, especially deriving from personal or private causes. Also, officials may not have a criminal record, nor discriminate against or violate personal rights. They have to be politically neutral with regard to dealing with governmental actors, other organisations and groups, and they may be sanctioned for disrespecting fiduciary duties or confidentiality agreements. In addition, officials may not accept gifts or benefits, bribes, commissions or other promises. Finally, they are not permitted to participate in betting activities (Asian Football Confederation, 2011c: 3–7).

A crucial infringement mentioned in the Disciplinary Code is doping abuse. For the purpose of specifying respective violations and sanctions, AFC has produced the AFC Anti-Doping Regulations. Having accepted the World Anti-Doping Code of 2009, AFC as the major governing body of football in Asia is responsible for fighting against doping abuse and implementing doping controls. In order to ensure the ethical values of sports, to protect the physical and mental health of players as well as to guarantee a fair competition, AFC has installed the AFC Anti-Doping Unit and an advisory group responsible for therapeutic use exemptions. In accordance with the FIFA Anti-Doping Regulations, AFC's regulations apply to AFC itself, its member organisations, players, clubs, support personnel, match officials, as well as other officials involved in activities, matches and competitions organised through AFC. Amongst the violations that can be sanctioned are the presence of a prohibited substance, its use or attempted use, refusal to give or failing to submit a sample, the failure to provide whereabouts information and missing a test, possessing a prohibited substance, trafficking substances, as well as administering prohibited methods or substances to players. The scope of sanctions that the AFC Disciplinary Committee may impose on individuals includes generally a two-year period of ineligibility for prohibited substances and methods, unless exceptional circumstances occur that allow reducing or eliminating the period. Trafficking and administering, meanwhile, can be sanctioned with a minimum of four years up to lifetime ineligibility. The same applies to violations that include a minor. Where a player is capable of providing assistance to help discover or establish violations, the Disciplinary Committee may reduce a part of the period of ineligibility. Players who have been sanctioned are then prohibited from participating in any competition or activity organised by AFC, a member association, league or club, the IOC, the International Paralympic Committee (IPC) or other international federations. In addition to becoming ineligible, players can also be sanctioned with a financial fine which, however, does not affect their ineligibility. Where more than two team members have violated the regulations, the Disciplinary Committee may sanction the club or even the respective association by deducting points, imposing forfeit, excluding a team from a competition's final standing or imposing a fine (Asian Football Confederation, 2010b: 8–9, 12–14, 36, 39, 46, 48–49).

In addition to the above-mentioned regulations and codes, AFC has produced legal frameworks for various areas, such as competitions and leagues, coach education, Vision Asia and the Financial Assistance Programme (AFAP). These include, in particular, criteria for clubs participating in AFC competitions and the requirements that have to be fulfilled by the respective licensing bodies (member association/league). The requirements encompass aspects regarding management, technical standards, attendance, marketing and promotion, business scale, game operation, media, stadiums and clubs (Asian Football Confederation, 2008a, 2010c, 2010d, 2012a, 2010e, 2012b, 2011d).

Finance and marketing

The AFC Financial Report 2009 serves as the main basis for taking a closer look at the AFC's sources of income, its expenses and budget cycle plans. Having adopted the International Financial Reporting Standards (IFRS) as from 1 January 2008, in order to comply with the procedures defined by the Malaysian Accounting Board for all companies that are incorporated in Malaysia, AFC documents its financial statements in US dollars.

In comparison to the budgeted income for the year 2009, AFC recorded an income increase of 8 per cent (US$4.264 million) under consideration of the new commercial rights fees received from the WSG, while it reported a deficit of US$22.551 million. In accordance with the IFRS and following the advice of the external auditors Ernst and Young, the revenues deriving from selling the commercial rights for the period from 2013 until 2020 were, however, only to be recognised during the contract period succeeding each phase of the agreement's completion, so that recognition of the income of US$25 million from the WSG was delayed until the completion of the commercial agreement and thus treated as deferred income. Due to this revision, the revenue recorded was US$31.854 million, whereas it would have been US$56.854 million if the WSG income had been included. Expenditures, meanwhile, were reduced by 18 per cent from the approved budget of US$66.608 million to US$54.854 million. Against this backdrop the AFC Financial Report 2009 concludes that the deficit should be considered merely as a paper loss.

The income statement includes revenues from sponsorship and TV rights, FIFA grants, levies, fines and other sources, while the expenditure includes expenses for competitions, administration, development, events and manpower as well as other operating expenses. The most important source of income was the revenue of US$24 million generated through sponsorship and TV rights (75 per cent), followed by the grant of US$2.5 million received from FIFA. Furthermore, AFC received US$1,032,718 through the Vision Asia sponsorship,

US$617,118 through the Awards Night sponsorship, as well as US$3,156,981 through other sources such as levies and fines.

Of the total expenditure, meanwhile, 70 per cent was invested into competitions. AFC spent US$23,702,497 on men's club competitions, US$11,856,608 on men's national team tournaments, US$1,213,239 on youth formats, US$192,061 on Futsal, and US$1,141,620 on women's competitions. Moreover, a considerable amount of US$10,928,148 was spent on administration, US$3,200,476 on educational matters and US$2,221,520 on Vision Asia and the Financial Assistance Programme.

The balance sheet for the year 2009 states total assets of US$26,928,111 divided into non-current assets of US$7.847 million (including property, plant and equipment) and current assets of US$19.081 million (including receivables and amounts owned by FIFA). For the purpose of comparing the approved budget with the actual one, AFC has included the deferred income received from the WSG. Accordingly, the total revenue was measured at US$56.854 million in comparison to the budgeted amount of US$52.590 million. The unexpected increase derived from sponsorship money for Vision Asia and the Awards Night, as well as fines and levies. The expected expenditure of US$66.608 million was reduced to US$54.551 million, explained through having reduced the expenses for competitions by US$2.233 million, for administration by US$4.89 million, for Vision Asia and education by US$3.06 million, and for capital expenditure and financial assistance by US$1.871 million.

It is finally noteworthy that AFC is exempted from taxation in accordance with respective Malaysian regulations (Asian Football Confederation, 2011e: 4–10, 23–24; Kuan, 2011).

With regard to AFC's marketing policy, focus is put on those countries that have clubs participating in the AFC Champions League in order to exploit the Asian market as efficiently as possible. In fact, the highest revenue stems from the AFC Champions League, followed by the World Cup qualifiers and then the Asian Cup. While the significance of the Champions League has increased rapidly, clubs used to regard it as rather unattractive due to high travel costs and low prize money. This perception has, however, changed since 2009 with the introduction of new regulations, appropriate hosting facilities, a new stadium strategy and an anthem, as well as higher prize money and subsidies for travel expenses.

Meanwhile, the Asian Cup, the flagship tournament for national teams, has been of vital importance for the AFC's marketing activities. Its 2011 edition, which took place in Qatar, constituted a platform for expanding and enhancing it as brand. According to AFC it was the most successful Asian Cup ever to be held, not least in terms of broadcasting and viewing figures with an outreach to more than 100 countries worldwide. In collaboration with the WSG and the Qatar LOC, promotional activities were implemented such as pre-tournament events and the installation of a fan zone as well as elaborate promotions in the stadium and during the matches.

Against this background, AFC's marketing vision aims at developing Asian football in order to make the Asian continent the most lucrative football market on the globe. In order to pursue this ambitious objective the mission is threefold: first, to improve the commercial value of AFC, its competitions and activities; second, to design marketing strategies and monitor respective activities in order to increase the revenue; and third, to build a structure that allows for reinvestment in developing football. For this purpose the AFC Marketing Department operates in seven areas. The first one is sales and licensing management, which includes selling TV and sponsorship rights as well as corporate rights and licensing sales. In this respect it is noteworthy that, according to the new contract, AFC can engage more actively in negotiations with the WSG and potential partners as from 2013, while rights deriving from the AFC's social responsibility and grassroots activities fall exclusively under the responsibility of AFC. The second area is concerned with account management, which basically encompasses the relations with licensees, suppliers and partners such as the WSG and Chelsea FC. Third, the creation and implementation of marketing guidelines and regulations as well as venue operations are taken care of within the scope of marketing operation management. Managing the broadcast services and contents, meanwhile, includes the production of technical materials, video clips and trailers, as well as AFC-specific content. In order to enhance business development management, footage sales have been put on the agenda, as well as developing corporate licensing and merchandising products and online games, while marketing communication management includes aspects of branding, design and market research. Finally, the area of marketing consultation and development management consists of offering consulting to the member associations, leagues and clubs, providing workshops and seminars, as well as internal consultation services.

In perspective, the AFC marketing policy aims at reaching a level that allows competing with European football, based on promoting and establishing genuine local football cultures across Asia through investing into areas European cannot access as easily. In 2011 the AFC Head of Marketing, Satoshi Saito, suggested that this could be achieved through employing a strategy of localisation and community attachment, and thereby win the hearts for local teams and create a local football identity. This does not necessarily mean that fans would stop supporting European teams but they might become more attached to the local teams. According to Saito it is thus important to indicate that Asian money should be invested into the Asian sports economy and the development of local football structures rather than into European football (Asian Football Confederation, 2012: 88–89; Saito, 2011; Kuan, 2011).

Competitions

The very core of AFC's *raison d'être*, planning, organising and implementing competitions and tournaments, as well as friendly matches, defines AFC's administrative and regulatory competence. This essential area of activity is not only based on the AFC's obligation and right to regulate and monitor compliance with the competition format, but also to control the bidding processes and election of a host country, as well as proper organisational conduct by the local actors.

The primary tournament organised by AFC is the Asian Cup. It was incepted in 1956 and has taken place ever since on a four-year basis. Its aim is to identify the best Asian national team, drawing upon a selection of teams that undergo a qualification phase. AFC has also installed continental youth versions of the competition, with the Under-19 tournament having first taken place in Kuala Lumpur in 1978 and the Under-16 version in 1984. The Women's Asian Cup, meanwhile, has been in place since 1975, followed by the Under-16 tournament in 1984. Both tournaments were first organised by the Asian Ladies Football Confederation before it dissolved in 1986 and the responsibility was transferred to AFC. Further competitions for Asian national teams that take place under the auspices of AFC are as follows: the FIFA World Cup qualification for Asia, the Olympic Games qualification for Asia, the AFC Challenge Cup, the AFC Under-22 Championship, the AFC Under-14 Championship, the AFC Futsal Championship, the AFC Under-19 Women's Championship, as well as the AFC Under-14 Girls Regional Championship (Weinberg, 2012: 544–545; Asian Football Confederation, 2004: 104–105; Duerden, 2011).

The other branch of AFC tournaments covers international matches between Asian clubs. It includes first and foremost the AFC Champions League, which constitutes the highest platform for Asian teams to compete on. As successor of the Asian Club Championship, which was founded in 1967, the AFC Champions League was relaunched and received a new format from the 2002/03 season, following the model of UEFA's counterpart. Including 32 teams, of which a selection from the top Asian leagues qualify directly, the AFC Champions League works through an assessment ranking system that measures the strength of a particular league based on defined indicators, including structure and financial status. Further AFC club competitions, meanwhile, include the AFC Cup, the AFC President's Cup and the AFC Futsal Club Championship. The President's Cup was introduced under Bin Hammam and is thought of as an opportunity for teams from less-developed leagues to develop and play at an international level. The AFC Cup and the President's Cup are viewed as very important and absolutely necessary because a lot of countries do not have the resources, facilities and structures to meet the requirements for the Champions League, but can participate nonetheless at an international level through these competitions, thereby providing clubs and players with the possibility of enhancing their performance. In 2011, AFC forecasted that five to eight countries participating in the AFC Cup might join the AFC Champions League in a few years' time, for example Vietnam, India, Jordan, Kuwait, Oman, Singapore or Thailand, while countries such as Myanmar are eager to be promoted to the AFC Cup. This logic is thought of as transparent system that motivates the clubs and leagues to develop and upgrade (Weinberg, 2012: 545; Asian Football Confederation, 2011f, 2011g, 2011h, 2011i; Tokuaki, 2011).

In terms of financing and enhancing the commercial output of its competitions, AFC has established several partnerships. Most important is the contract running with the World Sport Group according to which AFC will receive a total of US$1 billion until 2020. Also, AFC finalised sponsorship deals via WSG with Qtel, a telecommunication company from Qatar, and Emirates. The deal with WSG covers the transfer of marketing rights including client service, management, TV broadcasting and sponsoring. In terms of implementation, however ,AFC works together with WSG in relation to branding and operational matters. For a Champions League match, for instance, WSG would assign a marketing and TV manager, while AFC would assign a local general coordinator and media officer who consult with the match commissioner; after the match, WSG and AFC would then hand in a report. At match sites AFC would also provide staff in order to monitor the activities. When organising workshops or seminars, AFC meets with WSG to exchange relevant matters as well as general marketing issues. As for the appointment of match staff, AFC would ask the member association to nominate two to four general coordinators depending on the country. Afterwards, AFC invites the nominees to an assessment meeting and a workshop, followed by a monitoring procedure and an evaluation. The appointed match commissioners, meanwhile, are responsible for matters such as security, while the general coordinator and the media officer are rather concerned with operational matters such as the proper kick-off time, match organisation and media facilities. Overall, there are around 100 match commissioners in Asia who have obtained a qualification through a four-day course (Weinberg, 2012: 545; Tokuaki, 2011; Asian Football Confederation, 2009a; Glendinning, 2008; Emirates, 2009).

Another important aspect of organising and coordinating competitions is the international match calendar. Basically, FIFA decides its match days, for example for qualifying tournaments, but at times AFC would ask FIFA to provide alternative dates because the scheduling may have been in line with UEFA's dates but not with AFC's. The difficulty stems from the fact that in Europe teams sometimes play for instance on Fridays so that they can play again on the following Tuesday, while that procedure would not work in Asia, where the time difference for instance between Saudi Arabia and Australia makes it impossible to play on a Friday and again on a Tuesday. It is furthermore important to coordinate with the national associations, because in the East the season runs mostly from March until November while in the West it is from September until May, so fixing the dates for AFC proves to be a challenge. Scheduling matches in August, for example, is difficult because the western countries would refuse that period while the eastern countries would conversely be in favour of it. The same applies to the months of December and January, when in some countries the weather conditions may impede a smooth conduct. Also, it has to be taken into consideration that religious rituals such as Ramadan in Muslim societies affect the planning. Thus, AFC has to collect all relevant information on the leagues' dates and identify vacant periods of time. Accordingly, a proposal would be created, sent to the member associations, then revised and finalised. Requiring a long-term perspective, it is necessary for AFC to ensure the planning of the match calendar for the next few years to come. In

2011, for instance, the calendar until 2014 was already agreed upon. In this respect FIFA has been generally very receptive of AFC's priorities and challenges, while other confederations might have opposing interests (Tokuaki, 2011; Asian Football Confederation, 2011j, 2011k).

In terms of organisation, the Asian Cup is taken care of by a separate sub-department, together with the responsible committee. For the Asian Cup in 2011, approximately 120 employees were involved in the implementation process. Since it is the AFC's flagship tournament, however, the implementation requires cross-departmental cooperation. At the beginning of the planning process, all tasks would mainly be undertaken by the Asian Cup Department, including designated managers for media, marketing and ticketing. The department is responsible for communicating relevant issues with the Local Organising Committee for a period of four years and it conducts frequent checks on logistical matters, infrastructure, stadiums, hotels, training facilities and media requirements. About six months in advance of the tournament, the entire Competitions Department then engages in setting up the actual tournament structure, and other relevant departments become involved as well. It is noteworthy that the Asian Cup's and other competitions' regulations basically accord with FIFA's requirements with regard to system and structure, while some specifications may vary, e.g. the capacity of the stadiums (Asian Football Confederation, no year-a; Tokuaki, 2011).

The focus of the AFC's competition policy, however, has been set on the AFC Champions League as being the primary commercial product. For the purpose of enhancing its significance, AFC created a project team in 2006 (with the help of the Japanese Football Association) in order to visit member associations within a period of two years to examine structures and assess the quality of the leagues. The team travelled to 21 countries and defined criteria before narrowing down the selection to 17 and eventually picking 11 countries. Moreover, the project included the improvement of the commercial value of the Champions League, its brand identity and its visibility. In 2009 the competition's budget was increased from US$5 million to US$20 million, including prize money and performance bonuses. Ever since, media interest has increased as has as the number of spectators. Nonetheless, further improvements are necessary and therefore AFC continues to conduct regular visits by special mission teams, including staff from all departments, in order to optimise the levels of infrastructure, administration, finance, media, marketing and education. Despite the positive developments in the last few years, some leagues continue to struggle with meeting the requirements. Thus it may be that participating countries might actually be excluded. In the meantime, AFC encourages member associations participating in the second-tier competition, the AFC Cup, to raise their standards so that ultimately the goal can be achieved of having more leagues participating in the AFC Champions League.

With regard to the media dominance of European leagues being broadcast across Asia, the AFC's Head of Competitions, Suzuki Tokuaki (2011), expressed his belief that it does not necessarily pose a threat to the popularity of Asian football. On the contrary, it could help enhance the general popularity of football in Asia and, if AFC competitions reach a higher standard, create a spill-over. For the purpose of increasing the potential of Asian leagues, AFC has designed an additional initiative, the so-called Professional Football Development Project. Formerly a separate programme, it has been connected with the Vision Asia programme. It is a comprehensive programme aimed at professionalising structures through tailor-made instruments. Working together with FIFA on a country-by-country basis, AFC has approached medium-level leagues, such as in Jordan and Myanmar. If a country asks AFC to assist, respective services would be provided. But even countries such as South Korea require regular inspection and assistance; in fact, the only country having fulfilled all criteria was Japan in 2011.

The ultimate target of the approach is to advance the performance of the national teams, which is of course connected to the league system, its structure and the local level of play. Presumably this can only be achieved through taking the cultural differences into account, making thorough assessments and offering help according to the specific needs, and resolving the particular issues through organising workshops including all directors and relevant officials from the associations, the leagues and the clubs. Furthermore, it is regarded as inevitable to collaborate with governmental actors, since AFC requires all clubs to be commercial entities and in some countries they are registered as societies falling under the competence of the ministries responsible for sport. Thus AFC discusses directly with governmental actors or even urges the media to put on pressure so that the legal framework is changed and the clubs fall under, for instance, the ministry responsible for commerce. Also, it is at times necessary to engage in dialogue on stadium matters because in some countries the stadiums do not belong to the clubs and the facilities cannot be transferred without the support from the government; this was particularly a problem in the Middle East. With regard to issues of corruption and match fixing, meanwhile, AFC members such as the Chinese and the South Korean associations have met with governmental actors to discuss solutions, in the context of which AFC has merely participated as observer.

In relation to the perspectives and potential of AFC's competitions, Tokuaki stated that there was still a big gap between Europe and Asia, but that the latest developments have shown how it decreases and how the Asian top leagues can catch up, for example in Japan. Anyhow, it is vital that the majority of Asian countries draw level with the better ones, because at the end of the day, all play on one continent. Given the recent developments in the Middle East and the decision that Qatar hosts the World Cup 2022, Asian football as a whole and its national teams' performances could experience a significant enhancement. Against this background, AFC works closely together with the member associations to create a road map. Also, the hosting of the Asian Cup 2011 in Qatar has shown that in terms of infrastructure and facilities, the highest standards have been reached. Interestingly, AFC has engaged in discussions with UEFA not only with regard to learning from each others' experiences but also regarding if and how to create mutual competitions or exchange programmes, bearing in mind that the Asian market is very lucrative for UEFA (Tokuaki, 2011).

Technical and educational matters

Under consideration of the challenges Asian member associations face in terms of developing football, AFC has identified educational measures in the technical sector as one of its key activity areas. In this sense, AFC has established itself as the confederation that concentrates most on development and education through direct implementation. For this purpose it was deemed necessary not only to establish a transparent licensing system based on UEFA's model (A, B, C and the professional licence) but to foster top-level education, namely the qualification of elite instructors through respective courses and elite education seminars. In 2005 AFC held the so-called Year of Instructors, in the course of which an instructor panel was created with the minimum requirement of holding a coaching A-licence. Later, a professional licence was required and the panel reduced to ten members. Similar panels were also introduced for goalkeeping, conditioning, disability football and refereeing. Prior to this the member associations used to select instructors themselves; now AFC sends its experts to make an assessment and a selection. The main idea behind it is to conduct as many train-the-trainers courses as possible in order to create multipliers, especially in big countries such as China and India. The same approach relates to expanding knowledge on competition organisation, media relations and marketing, as well as sports medicine. In the grassroots area, AFC conducts regional events including C-licence courses and refereeing courses, during which the best participants are selected and invited to courses and seminars taking place within the framework of Project Future. The introduction and expansion of youth tournaments throughout the last few years has added to this development (Asian Football Confederation, no year-b, 2005, 2010e, 2010f, 2011l, 2011m; Al Sabah, 2011).

The diversity and different needs of the AFC's member associations, however, constitute a challenge to delivery of AFC's education policy. Most countries are still operating at a very low level and the outcomes of the AFC's initiatives so far have not been evaluated. From the perspective of both AFC and the member associations, some aspects of the programmes show a lack of know-how and proper assessment. Further improvement is necessary, especially with regard to making efficient use of resources and creating national development plans or strategies. Some initiatives have been too hasty and do not really aim at ownership and sustainability because some member associations rely too much on external support. Therefore, basically three conditions have to be fulfilled: first, governments, ministries and the education sector have to be supportive; second, the member associations have to take responsibility; and third, AFC has to function as a proactive confederation. The problem, however, is that the focus is mostly put on developing elite structures, whereas AFC aims at putting more emphasis on the importance of developing the game at the grassroots level, which in turn requires a long-term plan. Therefore there has to be a more systematic approach including an adequate monitoring system so that, if member associations do not perform, they will not receive any more funding. This implies potential political conflicts. In fact, one might find some officials of member associations who would not be interested in developing their own capacities and prefer relying on private academies or the AFC's initiatives, for example in some West Asian countries where this philosophy has not, however, helped them become more successful. Meanwhile, other countries have indeed done some impressive work, such as Japan where the installation of regional centres and collaboration with communities have affected the grassroots level positively in terms of higher participation and more opportunities for young players to perform in competitive formats. In this sense, the AFC Festivals of Football form an important platform for young players to compete at an international level while getting to know travelling, living in hotels and meeting other cultures. Also, the implementation of coaching, refereeing and even physiotherapy courses in the context of the Festivals has helped raise awareness amongst member associations of the importance of fostering all aspects of grassroots and youth football (Whittle, 2011; Al Sabah, 2011).

With regard to the delivery of the AFC's grassroots policies within the Vision Asia programme, deficits have been identified. While the administrative and organisational dimensions run well, the technical side often does not meet the desired objectives. Mostly being implemented at city and provincial levels, the national associations are hardly actively involved in the schemes. But if one is determined to work out a national development plan or coherent grassroots policy, all actors are needed. In this respect, AFC might have to engage more at the national level in a consultative role in order to ensure cohesiveness. Referred to as positive example is Thailand, where the AFC's Vision Asia scheme encompasses the national and the city levels through constant collaboration amongst the local actors with the aim of producing sustainable outcomes and change. In this scenario, AFC has shown that being a proactive confederation does not exclude the member associations; on the contrary, it means that AFC gets the ball rolling together with its partners so that at the end of the day the member association is the main stakeholder. In this sense, AFC might also have to become a bit more flexible and open-minded regarding working together, for example with FIFA, to eventually meet the needs of a particular member and ensure progress through an elaborate monitoring and evaluation system, as well as coordinate the schemes and dates with regard to management, technical matters and marketing to avoid redundant work. Also, experience has shown that a one-size-fits-all approach is obsolete on a continent like Asia. Since synchronisation and coordination efforts are complex and difficult, resources have to be managed efficiently and flexibly according to the needs of the particular country. This requires setting up a framework and environment including all actors and creating sustainable links between clubs and schools in order to provide young players with perspectives and foreseeable pathways. Thus advanced leagues can assist by sharing their knowledge and expertise on fostering youth development and grassroots schemes without imposing their model on other countries (Whittle, 2011; Asian Football Confederation, 2006, 2007, 2011n; Flath, 2012).

Meanwhile, with regard to the area of refereeing, the quality and quantity of highly qualified personnel is one AFC has been putting more emphasis on. By analogy with the problems identified in coach education and grassroots football, the major problem is that there is a large discrepancy between high- and low-level refereeing performances in Asia due to the disparity amongst the member associations. Accordingly, appointments for international matches have been difficult at times, while the number of FIFA referees has in fact increased in the last few decades. Another problem is the scale of corruption, match fixing and bribery, which can only be solved through professionalising the game and educating the referees properly. With this in mind, the major aim of AFC is to work on decreasing the discrepancy between high- and low-level countries through conducting courses and seminars. In particular, Project Future is regarded as an effective initiative. Again, problems may occur when member associations do not appreciate the AFC's initiatives. Generally, however, the educational schemes are perceived as good. Of further importance are not only the technical education but the professionalisation of administrative structures and the respective installation of refereeing departments within the member associations. And with regard to the problems of match fixing and bribery, the role of AFC as a decisive stakeholder is viewed as very important in terms of exchanging experiences and working together. Finally, it is thought of as vital to constantly reassess the current situation, monitor progress and think in the long-term in order to narrow down disparity between the associations (Kashihara, 2011).

With respect to the development of women's football in Asia, AFC has committed itself through the establishment of a respective department. As in other regions in the world, women's football is a growing sport but still requires a lot of groundwork, persistence and long-term commitment. Most importantly, facilities are needed since more than often the men's teams take up most of the available facilities. A precondition for changing these circumstances is of course a mindset according to which the women's game receives more attention, and the necessary acceptance amongst players, parents, coaches, clubs and officials. As a matter of fact, recent developments have shown that more and more girls are interested in and passionate about the game, which is advantageous in order to create more opportunities for them to play and compete. At the elite level, meanwhile, Asia has strong teams with senior and youth world champions amongst them. At the FIFA Women's World Cup Conference in 2011, the Technical Study Group indicated that Asian teams are very strong and have improved in terms of technical ability, speed and decision-making. Also, the fact that the IFAB has decided to allow Muslim players to wear the hijab will probably increase interest and participation in the game. With successful teams such as Japan, DPR Korea, South Korean, Australia and China representing the Asian continent, AFC has great interest in further developing women's football. The establishment of youth tournaments and competitions serves as expression of this agenda. Although the AFC Women's Department is primarily responsible for coordinating the competitions, it does work closely together with the technical division in order to produce more female coaches and referees, as well as respective grassroots structures. In this sense the women's competitions work along relatively strict criteria requiring member associations to have a sufficient number of local women's teams and competitions, as well as qualified coaches. In addition, a team has to appoint at least one female coach in order to provide the opportunity for them to improve and gain more experience. Finally in the context of the AFC's development schemes, more resources have been made available for courses for female coaches and referees, as well as media staff and sports medicine experts (Teo, 2011; Asian Football Confederation, 2011o, 2011p).

Another interesting arena AFC has become active in is Futsal. Having gained increasing popularity throughout the last few years, AFC has decided to push Futsal forward through the installation of respective educational measures. The Futsal development scheme mainly contains coach education courses as well as the establishment of a panel with 12 instructors. In addition, elite education seminars have been brought on the way, as have regional instructor courses and a licensing system with three levels covering Futsal-specific content, as well as technical, tactical and conditioning aspects. In this respect AFC has been more progressive than other confederations. Even UEFA expressed interest in learning from AFC in the area of Futsal. For the sake of development, however, further standardisation and recognition procedures are necessary. In 2011 AFC counted a total of almost 650 qualified elite personnel throughout Asia including coaches, referees, administrators and match commissioners. If a member association does not have an elite instructor, AFC would send one and cover the costs for travel and accommodation. Supporting the member associations with their plans, AFC would, however, leave it to them to decide on how many courses they want to conduct. Iran, for example, requested 30 courses for 2012, while Bahrain only needed two. Another major area is the development of strategy plans for countries, which previously used to be incorporated into the Vision Asia programme. For this purpose AFC sends mission teams in order to make assessment and create a road map. Two pilot projects were brought on the way in Iran in 2009. AFC has an officer who is in charge of Futsal development programmes for member associations, with a strong focus on improving the national league and competition level. In this context, strategies have been produced with regard to club development and professionalising the Futsal structures, both in administrative and technical terms. The strategy includes encouraging football clubs to incorporate a Futsal branch, while in some countries there are exclusive Futsal clubs that still have deficits in terms of youth and grassroots development. Taking into consideration that technical skills are also vital for large pitch matches, AFC expects normal clubs and the associations to show more interest in Futsal in future. Working together not only with other AFC departments but also FIFA, the Futsal Department is confident about its growing importance. Therefore, one of the major challenges is to enhance the popularity of Futsal and expand its visibility and media presence (Targholizade, 2011; Asian Football Confederation, 2011q, 2011r).

Development: Vision Asia and the Financial Assistance Programme

The Vision Asia programme is not only linked to but in fact constitutes the centrepiece of the AFC's educational, technical and development activities. It was initiated during the early days of Bin Hammam's presidency in 2003 as an incentive to convince 'Asia's football loving nations to recognise the importance of developing football from the sporting, business and social perspectives', with the overall objective of 'elevating football in Asia to the front rank of world football'. The concept of Vision Asia grounds on a systemic approach to reduce discrepancies not only between Asian countries themselves but also other regions by means of employing tools that focus on both 'on-field and off-field elements of football'. It is the conviction of AFC that 'sustained and integrated development programmes will drive Asian football to glory'.

Given the heterogeneity of the Asian continent, Vision Asia has been designed under consideration of the distinct economic, social and political settings of the particular countries taking part in the programme. It is thought of as providing 'solutions in a flexible and modular manner to cater to the individual needs and capacities of member associations'. By analogy with the number of players of a team on the football pitch, the strategic framework of Vision Asia builds on 11 elements which have been defined as crucial for the advancement of Asian football: member associations, marketing, youth, coaches, referees, sport medicine, women's football, competitions, Futsal, media and fans. According to the concept, these elements become relevant at all levels of organised football: the district level (school, amateur), the provincial level (amateur, semi-pro) and the national level (professional).

As for the member associations, the primary objective is to enhance their administrative and operational structures. With regard to the marketing element, the idea is to provide expertise for increasing the commercial value and generating more income. Meanwhile, the enhancement of grassroots and youth football is based on trying to increase the number of participating boys and girls, as well as creating appropriate facilities for this purpose. The establishment of coach education schemes, equipping educators, coaches and instructors with the necessary qualification, is thought of as one of the most important factors in raising the standard of play. The same applies to referees and sport medicine specialists. Women's football and Futsal shall be developed in their entirety with a focus, however, on creating and improving elite national competitions aimed at helping high-quality teams and clubs to emerge. The last two elements are concerned with professionalising the media structures and recognising the role of the fans, for both are viewed as essential components with regard to popularising the game (incl. quote Asian Football Confederation, no year-c: 1–5; 2003; 2008a; Weinberg, 2012: 546–547).

Vision Asia is basically a combination of consultation services and educational courses. In the actual delivery of its policies, Vision Asia has been de facto reduced from 11 elements to five phases including five to six elements.

First, focus would be set on administrative matters, especially at the city and provincial levels. This includes the creation of an organisational structure, the improvement of the Statutes as well as staff-related aspects based on an assessment in order to identify the deficits. Accordingly, an initial road map would be produced followed by a proposal the association can conduct a crosscheck on. The next step would be the planning and implementation of a workshop in which AFC officials share their knowledge and expertise before finalising the road map, which for instance could first include the restructuring of the congress, then the committees and finally the staff. If a particular association opines that the recommendations cannot be implemented, a compromise would be made with a follow-up meeting, and at times other AFC departments such as the Legal or the Grassroots and Youth Department may be consulted, while generally the main competence lies with the Vision Asia personnel unless there is a major issue that has to be solved. It is noteworthy that some projects have indeed been suspended and not continued, such as in Oman, Bahrain and Uzebkistan, while the city projects for instance in China have been in place basically from the very beginning of Vision Asia. Also, the projects in India have been going on for a considerable duration, including locations such as Manipur, Kerala, West Bengal and Goa. The same applies to the first-generation projects in Vietnam and Iran. Amongst second-generation countries are Bangladesh, Thailand, Sri Lanka and the Philippines. Except for Sri Lanka, Bangladesh and Pakistan, all projects are city projects. With regard to the monitoring and reporting procedures, AFC requires annual meetings and respective presentations.

The second phase, meanwhile, is devoted to amateur and club development in connection with coaching and refereeing courses. The third phase concentrates on the school leagues, in the course of which the Vision Asia Department cooperates with the Grassroots and Youth Department, which assigns one officer to join the Vision Asia delegation to make an assessment, create a road map and plan a workshop before the school league is relaunched, based on prior educational courses in cooperation with the AFC's Coach Education and Refereeing staff. In the meantime the Vision Asia Department functions as the project management entity with a coordinating role (Sugihara, 2011; Asian Football Confederation, 2011s, 2011t). (The fourth and fifth phases – in principle concerned with Futsal, women's football, marketing, media and fans – have been put on hold.)

Under consideration of the cultural context and the educational and political systems, all Vision Asia projects are based on close cooperation with local actors including district, state and national associations, clubs, governments and ministries, Olympic councils, the commercial sector, media and fan representatives. At times, the perspective of governmental actors may constitute a challenge, but their support is considered essential since AFC does not implement the measures but rather assists in doing so, and therefore it is regarded as necessary that all local actors commit to the initiatives. Due to the understanding of Vision Asia as a consultation service that depends on the willingness of the local actors, the mere provision of know-how and assistance is not sufficient. Hence the success of a project very much depends on the degree of commitment and the leadership skills of individuals involved in the project. In cases in which the partners have had a clear direction and ambition to initiate change, the working processes have been perceived as more efficient. While developing the relevant structures is considered in principle not very difficult, disadvantageous circumstances in terms of local management may be a hindrance. As pointed out by the Acting Head of the Vision Asia Department, the local key players have to be genuinely interested in improvement and optimisation, while not necessarily having to be in possession of in-depth technical knowledge but rather being dedicated in terms of appointing and managing competent experts and staff. Also, it has become clear from an AFC point of view that mentalities and mindsets play an important role as to how the projects are implemented. While it can be useful to draw upon the experiences of well-advanced countries such as Japan, it is thought to be crucial to bear in mind that copying a concept does not do the job. A positive example referred to is Iran, where the federation's President plays a leading role. Interestingly, other countries exist which are not part of the Vision Asia programme but where a spill-over effect has led officials to take similar initiatives themselves. Another good example is supposedly Thailand, where the Vision Asia project has undergone a solid development due to the strong leadership of the General Secretary of a club who founded a provincial federation, became its General Secretary and proved to be a good manager, knowing how to recruit and delegate tasks (Sugihara, 2011).

In financial terms, meanwhile, AFC covers the travel and lodging costs for its staff. The courses are also paid for by AFC, but the expenses for local staff and the infrastructure have to be paid for by the association. When asked in 2011, the Acting Head of the Vision Asia Department pointed out that in future a new approach would encompass more coordination with the AFC's Professional Football Project and FIFA initiatives in order to complement the different programmes and avoid redundancies. While there was no detailed plan yet, he expressed his belief that the concept of uniting technical and administrative aspects in the programme is correct and necessary. Although there had been sporadic problems with the implementation, the basis had been formed for stepping up and using its potential for working more closely with FIFA in relation to its new Performance Programme, creating synergies. While FIFA would put more focus on the professional level as well as its infrastructure through the Goal Programme, AFC would concentrate on the amateur level. Appreciating the general concept of Vision Asia, which in fact has initiated a trend on how to tackle football development issues, he finally mentioned that the programme lacks financial resources (at the time only Chelsea FC was a sponsor), which, however, does not affect the actual consultation service so that all applications so far have been catered for (Sugihara, 2011).

The second major development programme of AFC is its Financial Assistance Programme. Formerly named AID27, it was created as the AFC's counterpart to FIFA's Financial Assistance Programme, aimed at supporting selected member associations with funds in order to build-up structures and invest in development. Associations with low economic growth and limited resources thereby received the opportunity to apply for funds, particularly for grassroots and youth football, women's football and Futsal. From 2008 until 2012, AFC assisted a total of 29 member associations including countries such as Afghanistan, Bahrain, Cambodia and Guam. The scope of the assistance encompassed the employment of coaches (maximum US$5000 per month), the implementation of coaching courses (maximum US$12,000 per year), other technical initiatives (maximum US$24,000 a year) and the development of women's football (maximum US$24,000 a year).

While the coach employment scheme was intended for coaches of both men's and women's teams – from Under-13 to senior levels – as well as for coaches of Futsal national teams, coaching courses were financed for C, B, A and professional licences as well as instructor and refresher courses. Other technical initiatives, meanwhile, included payments for administrators, for participation of teams in the Festivals of Football and other youth competitions, for refereeing courses, and for necessary equipment. For the development of women's football, the beneficiaries received funds for financing female coaches, girls' teams participating in the Festivals of Football and other competitions, and courses for female coaches and referees (Weinberg, 2012: 547; Asian Football Confederation, 2010: 5–9, 12, 15; 2011u).

In 2012 the programme was restructured and adjusted. Accordingly, it aims at enhancing the capabilities of the member associations through assisting with the employment of coaches, participation in youth competitions and the Festivals of Football, recruitment of qualified administrators, and the development of women's football as well as general infrastructure. Running until 2015, the recipients receive an annual grant which is defined by the AFC Executive Committee. Thirty per cent of the funding shall be invested into the employment of men's coaches and administrators, 15 per cent into the development of women's football and 55 per cent into other technical initiatives. In terms of monitoring the programme, the AFC's responsible committee has the right to implement annual audits as well as to evaluate the progress through a consultant, while a member association is obliged to initiate an audit through an independent auditor. In addition, a member association receiving funds has to employ a full-time administrator who manages the programme. Regarding the areas supported by the AFAP, a member association can specifically apply for grants for the employment of men's national and regional team coaches of Under-13 to Under-23 teams, Futsal and beach soccer. In order to develop the women's game, AFC provides financial resources for the employment of female coaches for Under-13 to regional and national, for hiring female administrators, for taking part in the AFC's competitions for national teams and the Festivals of Football, for courses designed for female coaches and referees, for the organisation of competitions, for the organisation of national, regional and grassroots activities, as well as for the hosting of competitions. The scope of technical initiatives supported by AFC, meanwhile, includes the provision of funds for administrative staff and additional coaches, the participation of youth national teams in AFC competitions, the organisation and participation in coaching and refereeing courses, as well as investments in equipment. In case the grants are misused, the AFAP Committee has the right to suspend further payments, demand a payback or take other actions that are considered necessary (Asian Football Confederation, 2011v; 2012d; 2012b: 4–6).

Social responsibility: Dream Asia

In the last two decades, sport-related companies and organisations have come to realise the importance of promoting social responsibility activities. Joining the trend of either utilising sport as a tool or contributing to the improvement of sport infrastructures, AFC has created its social responsibility concept named 'Dream Asia', which aims at building upon the supposed effectiveness and popularity of football, specifically with regard to achieving wider development objectives such as poverty and hunger reduction. In this context, AFC has identified several areas in which football could be used as a tool to create awareness and address social issues. The themes covered include social inclusion and gender equity, fair play and sportsmanship, children's rights, conflict prevention, and health promotion (Asian Football Confederation, 2008b, 2010h).

The establishment of the AFC Social Responsibility Department has been based on the idea of, first, becoming active within the organisation and the football sector itself through organising and facilitating events and activities for staff, players and coaches in order to enhance the credibility of AFC as a socially responsible actor; and second, making a name for itself through community outreach initiatives aimed at contributing to social development goals. Although its financial resources are rather limited in comparison to the AFC's core activity areas, the value of promoting social responsibility in and through football across Asia is regarded as high. It is thought of as chance to draw attention to social problems by means of encouraging, for instance, players or coaches to become active and use their popularity to give something back to society. For this purpose, the responsible department has been looking into how to connect the Festivals of Football with social responsibility events and incorporate the local communities. For 2012, for instance, the idea was to conduct events through which coaches would learn how to create awareness of social issues and community work. Furthermore, information events and education sessions would become part of the Festivals, in the course of which people from the local community can participate and receive, for instance, a medical check-up for free, while the head coaches would give training sessions to children and organise small matches (Ra, 2011).

Another approach is the Midnight Football Programme, which is viewed as explicitly addressing social issues in collaboration with local actors including the football associations. It aims at being driven by the community, based on ownership and sustainability. In this sense AFC sees itself as gate opener that can bring in resources, such as coaches, referees and others, while relying on the commitment of the other actors. Educators and doctors can also be involved, thereby linking football with social community work. While AFC can provide financial support in the early phase, the objective is to follow a long-term approach in which AFC slowly fades out its involvement and transfers ownership. The pilot project, which took place in Kuala Lumpur in 2011, included the local government providing human and financial resources as well as participating in direct operational matters. Within the project, the community work aspect as well as the collaboration with the local actors proved to be crucial for its successful outcome. Based on this experience, governmental actors should therefore be interested not only in contributing financially but also providing resources and knowledge, and in drawing upon the existing grassroots infrastructure. In Malaysia, the Department of National Unity has been committed to creating integrative programmes and it has recognised the benefits of the pilot project with regard to education and integration. Having also worked together with the local integration office in Kuala Lumpur, AFC expressed its hope to expand the programme on a national scale (Asian Football Confederation, 2011w: 16ff.; 2011x).

In terms of creating partnerships, AFC has coupled with the Food and Agricultural Organisation (FAO) of the United Nations, which AFC helps by raising awareness and providing funding, for instance for tree planting initiatives. With the ambition of positioning AFC as one of the leading sport organisations focusing on social responsibility, the partnership with the FAO is regarded as crucial. Problems of poverty, AIDS and hunger are very persistent in Asia and, due to the fact that FIFA has put its focus mostly on Africa, it gives AFC the chance of drawing attention to the Asian continent, not least in light of the World Cup 2022. For this purpose, collaboration with UN agencies is considered highly important in order to raise awareness and generate funding and commitment through platforms such as the AFC's competitions. For instance, the benefits of the opening game of the Asian Cup 2011 were divided between AFC and the Local Organising Committee, and AFC decided to donate US$400,000 entirely to the FAO, which used it for projects such as in Thailand, where relatively small amounts of money are supposed to help with creating food supply capacities. AFC also encourages national associations to at least send players into these regions and give the local community a feeling of hope and positive belief, although this can be considered only a minimal engagement. But it can help raising awareness and perhaps prompt more donations and funds. Moreover, AFC aims at helping the national associations to develop a social responsibility strategy and make it part of their policy, which is certainly a challenge given that most are concerned with their main business. Small associations especially simply do not have the resources, but AFC stresses that essentially the whole football culture is based on the foundation of the pyramid, namely the grassroots and community level. Also, collaboration with local NGOs and an AFC grant labelled Programme Support of US$25,000 offer potential starting points for engagement. In fact, it is the priority of AFC to activate the associations, and in this context the Midnight Football Programme is referred to as positive example for building something up together with the associations and the communities. Particularly India and China will receive more attention in the next few years because of the socio-economic problems in these countries.

Another interesting concept AFC has been looking into derives from a collaborative effort with a South Korean university aimed at using football for character building and education. While the AFC's technical activities are manifold, they lack an explicit character building concept, for instance for coaches. This is something where the Social Responsibility Department aims at working closer together with the AFC's technical departments. Furthermore, options are explored in how far it is possible to draw more attention to social responsibility issues in the context of the AFC's competitions, for example through players visiting local projects or stadium campaigns with banners relating to poverty problems or AIDS prevention. Another partnership AFC has entered is one with ROTA, a non-profit organisation based in Qatar under the auspices of the Qatar Foundation. With its focus on providing more and better access to education, ROTA works together with NGOs in 20 countries. AFC has considered this approach as a good basis for cooperation, and in 2011 a joint project was implemented in Doha. Against the backdrop of the World Cup taking place in Qatar in 2022, further possibilities are to be explored (Asian Football Confederation, 2009b, 2011y, 2011z, 2011aa, 2011ab).

With regard to promoting good governance, including rules of conduct, and tackling corruption and bribery both at the lowest and the highest levels, the Social Responsibility Department considers these to be very important but also delicate issues that have to be dealt with sensitively. The department sees its role in this context as having an eye on the developments, identifying the problems and raising awareness, while the whole process requires a lot of collaboration. Specifically, for instance, the department could initiate stadium campaigns but its capabilities are regarded relatively low. Drawing attention to social values and issues, however, even within the AFC headquarters itself, could create a culture that may have positive impacts in the long run. Further aspects that have become increasingly relevant are mental health issues of players as well as taboos amongst players, such as homosexuality. The department considers these issues to be important but perceives them to be rather secondary given that football in Asia faces other problems, and that developing structures and leagues are viewed as more urgent. AFC could again promote and address the social and ethical dimensions of football. Putting these on the agenda might indirectly contribute to diminishing negative trends, including hooliganism and violence. While being aware that football cannot solve all social problems, it is regarded as a tool to give people hope and perspective. Finally, as for finding sources of funding, the department is determined to address AFC's sponsors and get these more involved in collaboration with the Marketing Department. For this purpose, both local and international sponsors shall be approached for the smaller projects and the national associations respectively (Ra, 2011).

Interim conclusions

In order to assess the role of AFC, it is necessary to take a look at the specific policies and thematic areas it engages in. This includes an analysis of the strategies, programmes and outputs AFC puts forward.

The core competence of AFC is its task to run the game and ensure its jurisdiction and regulation. Incorporated into the system of world football as defined by FIFA, AFC legislates, implements and enforces regulations. These processes require not only collaboration with FIFA, but also negotiation with other actors including governments with regard to the legal context and the autonomy of the sport sector. The AFC Statutes are the most important legal text of the confederation. They provide information on the objectives, bodies, competitions, membership and financial matters. Frequent amendments, modifications and reforms have taken place in the context of FIFA's aim to harmonise the institutional framework, including aspects of players' transfers and anti-doping measures. The other considerable legislative texts of AFC include the Disciplinary Code and the Code of Ethics, both containing definitions of infringements and sanctions as well as the judicial bodies and particular procedures. Moreover, AFC operates Anti-Doping Regulations in accordance with the WADA Code, and it employs regulations with regard to activities such as the AFAP, Vision Asia and specific competition formats.

In terms of finance and marketing, AFC produces an annual financial report which contains an overview of the income, expenses and envisaged budget cycle. The major source of income is sponsorship and TV rights (75 per cent), followed by FIFA grants, levies and fines. It is notable that AFC has signed a marketing deal with WSG for the period from 2013 until 2020, generating a total income of US$1 billion. On the expenditure side, competitions make up 70 per cent, followed by administration costs, development and educational measures, events, manpower and operating costs. In terms of marketing, focus is put on the AFC Champions League, which produces the highest revenue. The ACL received a new format in 2009 and has been henceforth more lucrative for clubs to participate in. The Asian Cup, meanwhile, is the flagship tournament for national teams and has increased its viewership and outreach significantly in 2011. The overall objective of the AFC's marketing policy is to make Asia the most lucrative football market on the globe, to reinvest in grassroots structures and to create local football identities.

Competitions constitute an essential activity for the AFC. The planning, organisation and implementation of the tournaments are linked to the regulatory function of AFC. This also includes controlling the bidding processes and the work of the local organising committees. The primary competition for national teams is the Asian Cup. Respective youth tournaments have been established for boys and girls. Furthermore, AFC organises the World Cup qualifiers, the Olympic Games qualifiers and the Challenge Cup. For the clubs, the Champions League is the highest competition to participate in. It was already founded in 1967 and relaunched under the new name in 2002. Further club competitions include the AFC Cup and the President's Cup, both thought of as formats for clubs of weaker leagues to develop. In financial terms, marketing and sponsorship deals with WSG, QTel and Emirates reflect the growing commercial value of the AFC's competitions. This development is underlined by the fact that AFC employs special mission teams to improve and establish league structures. The establishment of the Professional Football Development Project – which includes working not only with member associations and governmental actors but also with other confederations such as UEFA in order to secure sustainable frameworks – adds to this trend. With regard to organising the tournaments, AFC collaborates with local actors and the WSG in order to appoint staff and ensure a successful implementation process. Moreover, producing the regional match calendar requires coordination with FIFA and may lead to difficulties due to the size of the Asian continent.

In the area of technical and educational matters, AFC puts a strong focus on developing the game at all levels through qualification schemes and assisting with direct implementation. The policy programme covers coach education, refereeing, management and administration, women's football, sport medicine and Futsal. The diversity and discrepancy between the member associations poses a major challenge in this context. Thus AFC depends strongly on the commitment and flexibility of the member associations, governmental actors and civil society, as well as FIFA. However, the aim is to activate the member associations as the main stakeholders.

AFC provides two development programmes, namely Vision Asia and the AFAP. Vision Asia constitutes the centrepiece of the educational activities and was incepted during the presidency of Bin Hammam. It contains 11 elements, encompassing most dimensions and all levels of organised football. The programme is designed as a combination between consulting services and educational courses. It requires collaboration and negotiation with various actors. The delivery of some of the projects has, however, turned out to be problematic at times due to lack of commitment.

Finally, AFC has produced a social responsibility policy named Dream Asia. It has two aims: first, the mobilisation of CSR-related activities within the organisation itself, and second, external programmes and initiatives. The overall idea is to utilise football as a tool and 'door opener' in relation to development issues such as health, hunger and community work. In this context, AFC partners with local actors and NGOs such as ROTA and the FAO. Themes such as inclusion and gender empowerment shall be put more on the agenda in order to raise awareness.

References

Al Sabah, Hasan (AFC Deputy General Secretary and Director of Education) interviewed by B. Weinberg, 7 October 2011, Kuala Lumpur.
Asian Football Confederation. (2003). *Vision Asia: Official brochure.*
Asian Football Confederation. (2004). *The power of dreams: 50 years of the Asian Football Confederation*, Kuala Lumpur.
Asian Football Confederation. (2005). *AFC Coach Education Department manual and policy.*
Asian Football Confederation. (2006). *Grassroots/youth handbook.*
Asian Football Confederation. (2007). *Grassroots and Youth Department presentation.*
Asian Football Confederation. (2008a). 'Regulations: AFC Asian Cup 2011: Qualifiers.'

Retrieved from www.the-afc.com/uploads/Documents/common/cms/afc/AsianCup2011 QualifiersRegulations20080606.pdf on 5 October 2011.

Asian Football Confederation. (2008a). *Vision Asia presentation.*

Asian Football Confederation. (2008b). 'AFC stresses social responsibility.' Retrieved from www.the-afc.com/en/component/content/article/20059 on 1 January 2008.

Asian Football Confederation. (2009a). 'AFC, WSG renew landmark partnership.' Retrieved from www.the-afc.com/en/media-releases/26117-afc-wsg-renew-landmark-partnership on 10 November 2009.

Asian Football Confederation. (2009b). 'AFC keen to help FAO fight hunger.' Retrieved from www.the-afc.com/en/component/content/article/24779-afc-keen-to-help-fao-fight-hunger on 1 December 2009.

Asian Football Confederation. (2010a). *AFC Disciplinary Code.*

Asian Football Confederation. (2010b). *AFC Anti-Doping Regulations.*

Asian Football Confederation. (2010c). *AFC club licensing manual.*

Asian Football Confederation. (2010d). *AFC Vision Asia club licensing regulations.*

Asian Football Confederation. (2010e). *AFC coach education policy and procedures for coaches.*

Asian Football Confederation. (2010f). *AFC coach education policy and procedures for coaching instructors.*

Asian Football Confederation. (2010g). *AFC AID27 regulations.*

Asian Football Confederation. (2010h). *Social responsibility takes centrestage.* Retrieved from www.the-afc.com/en/social-responsibility-news/31479-social-responsibility-takes-centrestage on 10 November 2010.

Asian Football Confederation. (2011a). 'About AFC.' Retrieved from www.the-afc.com/en/about-afc on 21 February 2011.

Asian Football Confederation. (2011b). *'Statutes of the Asian Football Confederation: Regulations governing the application of the Statutes and standing orders of the Congress,* 2011 edition. Retrieved from http://image.afcpresident.com/upload/library/AFC_Statutes_2011.pdf on 10 November 2010,

Asian Football Confederation. (2011c). *AFC Code of Ethics.*

Asian Football Confederation. (2011d). *AFC equipment regulations.*

Asian Football Confederation. (2011e). *AFC XXIVth Congress 2011: Financial report 2009.*

Asian Football Confederation. (2011f). *Champions Leagues assessment requirements.*

Asian Football Confederation. (2011g). *ACL general assessment report.*

Asian Football Confederation. (2011h). *Financial analysis of Asian professional leagues and clubs.*

Asian Football Confederation. (2011i). 'Call to improve AFC competitions.' Retrieved from www.the-afc.com/en/news/35851-call-to-improve-afc-competitions on 18 September 2011.

Asian Football Confederation. (2011j). 'AFC calendar of competitions 2013 (men).' Retrieved from http://images.the-afc.com/Documents/calendars/Comp_Calendar_2013_men.pdf on 5 October 2011.

Asian Football Confederation. (2011k). 'AFC calender of competitions 2014 (men).' Retrieved from http://images.the-afc.com/Documents/calendars/Comp_Calendar_2014_men.pdf on 5 October 2011.

Asian Football Confederation. (2011l). *Elite education seminar 2011 booklet.*

Asian Football Confederation. (2011m). *Education division and coaching education presentation.*

Asian Football Confederation. (2011n). 'From youth festival to World Cup.' Retrieved from www.the-afc.com/en/component/content/article/686-interview/35684-from-youth-festival-to-world-cup on 1 December 2011.

Asian Football Confederation. (2011o). 'MAs' roles in women's football.' Retrieved from www.the-afc.com/en/news/37414-mas-roles-in-womens-football on 12 December 2011.

Asian Football Confederation. (2011p). 'JFA chief sees positive women's trend.' Retrieved from www.the-afc.com/en/tournaments/women/afc-u19-womens-championship/36911-jfa-chief-sees-positive-womens-trend on 11 December 2011.

Asian Football Confederation. (2011q). 'AFC driving Futsal forward says Targholizade.' Retrieved from www.the-afc.com/en/tournaments/clubs/afc-Futsal-club-championship/35545-ali-targholizade-interview on 5 October 2011.

Asian Football Confederation. (2011r). *AFC Futsal presentation.*

Asian Football Confederation. (2011s). 'Vision Asia process.' Retrieved from http://images.the-afc.com/Documents/2011/vision-asia/Vision_Asia_Process.pdf on 25 July 2011.

Asian Football Confederation. (2011t). *Vision Asia implementation phases and key performance indicators presentation.*

Asian Football Confederation. (2011u). *AID27 presentation.*

Asian Football Confederation. (2011v). 'AID-27 Monitoring Committee meets.' Retrieved from www.the-afc.com/en/news/36630-aid-27-monitoring-committee-meets on 11 December 2011.

Asian Football Confederation. (2011w). *Dream Asia: Football's commitment to social development (brochure).*

Asian Football Confederation. (2011x). 'More Midnight Football boost.' Retrieved from www.the-afc.com/en/news/35340-more-midnight-football-boost on 16 July 2011.

Asian Football Confederation. (2011y). 'Twenty nations to receive AFC–FAO aid.' Retrieved from www.the-afc.com/en/programs-dream-asia-campaigns/campaigns-afc-fao/campaigns-afc-fao-news/36686-twenty-nations-to-receive-afc-fao-aid on 11 December 2011.

Asian Football Confederation. (2011z). 'AFC–FAO help 14,000 hungry people.' Retrieved from www.the-afc.com/en/programs-dream-asia-campaigns/campaigns-afc-fao/campaigns-afc-fao-news/36813-afc-fao-help-14000-hungry-people on 11 December 2011.

Asian Football Confederation. (2011aa). 'AFC signs MoU with ROTA.' Retrieved from www.the-afc.com/en/social-responsibility-news/32034-afc-signs-mou-with-rota on 7 January 2011.

Asian Football Confederation. (2011ab). 'ROTA keen on future cooperation.' Retrieved from www.the-afc.com/en/dream-asia-award/programs-dreamasia-awards-news/37291-rota-keen-on-future-cooperation on 12 December 2011.

Asian Football Confederation. (2012a). 'Fulfill club licensing criteria by 2013, AFC to MAs.' Retrieved from www.the-afc.com/en/events/elite-education-seminar-2012/37698-fulfill-club-licensing-criteria-by-2013-afc-to-mas on 16 February 2012.

Asian Football Confederation. (2012b). *AFC Financial Assistance Programme (AFAP) regulations.*

Asian Football Confederation. (2012c). *AFC Activity Report 2011.*

Asian Football Confederation. (2012d). 'AFC ExCo takes important decisions.' Retrieved from www.the-afc.com/en/news/38461-afc-exco-takes-important-decisions on 27 June 2012.

Asian Football Confederation. (no year-a). *Competitions department presentation.*
Asian Football Confederation. (no year-b). *The evolution of coach education in Asia presentation.*
Asian Football Confederation. (no year-c). *Vision Asia (brochure).*
Duerden, J. (2011). 'John Duerden: Asian Cup history.' Retrieved from http://soccernet. espn.go.com/columns/story/_/id/861355/john-duerden:-asian-cup-history?cc=4716 on 5 October 2011.
Emirates. (2009). 'Emirates remains committed to Asian football's future.' Retrieved from www.emirates.com/jp/japanese/about/news/news_detail.aspx?article=397114 on 10 November 2010.
FIFA. (2011). *Statutes.*
Flath, Thomas (AFC Director of Grassroots and Youth 2003–2006) interviewed by B. Weinberg, 13 February 2012, Mönchengladbach.
Glendinning, M. (2008). 'Qtel to sponsor 2011 AFC Asian Cup.' Retrieved from www.sportbusiness.com/news/168214/qtel-sponsor-2011-afc-asian-cup on 10 November 2010.
Johnson, James (AFC Director of International Relations and Development) interviewed by B. Weinberg, 11 October 2011, Kuala Lumpur.
Kashihara, Joji (Head of AFC Referees Department) interviewed by B. Weinberg, 23 September 2011, Kuala Lumpur.
Kuan, Bryan (Head of AFC Finance Department) interviewed by B. Weinberg, 30 September 2011, Kuala Lumpur.
Ra, Ingill (Head of AFC Social Responsibility Department) interviewed by B. Weinberg, 14 October 2011, Kuala Lumpur.
Saito, Satoshi (Head of AFC Marketing Department) interviewed by B. Weinberg, 10 October 2011, Kuala Lumpur.
Sugihara, Kaita (Acting Head of AFC Vision Asia Department) interviewed by B. Weinberg, 29 September 2011, Kuala Lumpur.
Targholizade, Ali (AFC Director of Futsal) interviewed by B. Weinberg, 30 September 2011, Kuala Lumpur.
Teo, Julie (Head of AFC Women's Department) interviewed by B. Weinberg, December 2011, email interview.
Tokuaki, Suzuki (AFC Director of Competitions) interviewed by B. Weinberg, 27 September 2011, Kuala Lumpur.
Weinberg, B. (2012). ' "The future is Asia"? The role of the Asian Football Confederation in the governance and development of football in Asia.' *The International Journal of the History of Sport, 29*(4), 535–552.
Whittle, John (Head of AFC Grassroots and Youth Department) interviewed by B. Weinberg, 11 October 2011, Kuala Lumpur.

8 Understanding the politics of football

Discussion and conclusions

Discussion

The role of AFC as a 'glocal player'

Examining sport politics and policies requires taking a close look at aspects of internationalisation, transnationalism and globalisation. While the latter term has been assessed through various disciplinary perspectives, with differing opinions regarding causal mechanisms, effects and reciprocities, the common denominator is that globalisation relates to the nexus between local, regional, national, transnational and international phenomena.

The debates on the effects and ramifications of globalisation have led to three broad streams. While the hyperglobalisers consider globalisation a unidirectional process that diminishes the power of nation states and creates a global society, the sceptics regard the supposed dissolution of the nation state and the homogenisation processes as a myth. The transformationalists, meanwhile, view globalisation as a unique order but as not entirely new, and they draw attention to its openness and discontinuity. In this context Robertson (1990: 26) has developed a five-phase model, which emphasises the temporal–historical path of globalisation.

Maguire (1994; 1999: 402; 2006: 356ff.) understands globalisation as a multidimensional and uncertain process characterised by transnational economic exchanges, communication and migration. He supposes that this leads to elaborate forms of interdependence and time–space compression. Corresponding with Robertson's model, Maguire has produced a five-phase sportisation model. With regard to the relation between heterogenisation and homogenisation, he has developed the notion of diminishing contrasts and increasing varieties, according to which cultures merge and the more powerful groups in society seek to control the processes. In contrast, other scholars (Amara *et al.*, 2005: 204) have argued that, in relation to football league systems, cases indicate an increase in contrasts and diminution in varieties. With respect to harmonisation effects, Houlihan (2009: 496) in turn draws attention to the example of international sport development regimes established by the international federations which, however, have varying impact on a domestic system subject to its size, power and success.

Another important definition and approach has been developed by Giulianotti and Robertson (2007: 168ff.; 2009: xi–xii, 5ff.). With regard to globalisation, they emphasise aspects of compression, global interconnectedness, connectivity and globality. Viewing football 'as a metric, a mirror, a motor, and a metaphor of globalization', they argue that globalisation cannot be understood as a process in which the western culture dominates others. In this sense they utilise the term 'glocalisation', which refers to how global and local driving forces interweave, synthesise or conflict with each other. With regard to club football, they mention the example of the hegemony of the rich western European clubs, while at an international level emerging football continents have been striving for more influence concerning the allocation of places for the World Cup. Moreover, Giulianotti and Robertson have identified five phases of a global history of football, of which the fourth and fifth are particularly relevant for this study. The fourth 'struggle-for-hegemony' phase ranges from the mid-1920s until the late 1960s and encompasses the hosting of the first World Cup, the professionalisation and global expansion of organised football, including the establishment of the continental governing bodies, as well as the political instrumentalisation of sports. The fifth 'uncertainty' phase, meanwhile, spans from the late 1960s until the early 2000s. It is characterised by the increasing political influence of non-European countries and the mass mediatisation of international football. Also, Giulianotti and Robertson mention that ever since the 1990s the international system of football has experienced an increase and diversification of actors, which are strongly associated with the rapid commercialisation of the game. In this context, Horne and Manzenreiter (2004: 4ff.) have pointed to the significance of the economic aspect of globalisation, considering for instance the commercial development of the World Cup and FIFA respectively as well as the trend to exploit the Asian market.

Building upon these insights, globalisation can be understood as a multi-directional time–space compression encompassing political, economic, social, cultural and historical dimensions, all of which revolve around tensions between global and local driving forces, homogenisation and heterogenisation. These assumptions have to be taken into account when assessing the role of AFC in Asian and global football. In particular, Maguire's and Houlihan's considerations on harmonisation and heterogenisation, as well as Giulianotti and Robertson's notion of glocalisation, serve as a useful basis for examining AFC's relations with and impacts on member associations, and the reciprocities between association football as originated in Europe and how it has developed in Asia. It also provides a framework for an understanding of how AFC has tried to promote a distinct Asian football culture, as well as how the regulation of football through governmental actors has affected football in Asia. In addition, the phase models by Maguire and Giulianotti and Robertson assist in the assessment of the historical development of football in Asia. Notably, the globalisation discourse is interrelated with the academic debate on the emergence of governance structures, including the development of international regimes, the relation between old and new institutions, and cultural political consequences.

Looking at the political, economic, social, cultural and historical context of football in Asia, the results show that the continent is characterised by a significant diversity and disparity. Asian countries' standards and conditions vary in terms of the level of play, the tension between popular interest in the national teams versus the local club teams, the leagues' structures and degree of professionalisation, commercialisation and mediatisation, the development of grassroots and youth football, the education of coaches, referees and administrators, the tension between regulation and autonomy, the interests of member associations and their individual representatives, the roles of governments and other actors, the status of women's football and its promotion, the issue of sport and religion, and the difficulties deriving from corruption and match fixing. The political, economic and socio-cultural circumstances have led to different developments of football structures resulting in the emergence of a core and periphery in terms of sporting and commercial success. Countries like Japan and South Korea have fostered professionalisation and commercialisation processes through implementing policies and programmes ever since the 1980s/1990s, thereby improving the levels of play and participation. Some countries, such as Hong Kong and Malaysia, have failed to maintain their status, while others continue to struggle and require fundamental measures in order to catch up with the advanced football nations. Meanwhile, economically emerging countries such as India and China have been commonly referred to as having great potential, but have not met the expectations yet. Countries from the Middle East, in turn, have increasingly invested more into their sport systems, while formerly strong nations like Iran and Saudi Arabia have experienced a setback in terms of success of their national teams. It is this very diverse setting within which to locate and understand AFC's perceptions, strategies and (inter-)actions.

The history of AFC dates back to 1954, when it was founded in Manila. Two aspects appear particularly interesting in this context. First, the fact that a Scot was significantly involved in the formation of the confederation indicates the interplay between a European influence within which organised football evolved at the regional level in Asia (in fact, the current working language of AFC is primarily English), and the desire to create a distinct Asian counterpart to UEFA and an organised voice for Asian associations at the global level. Second, the fact that in 1954 AFC consisted of merely 12 member associations indicates how great the expansion of the confederation has been in the last five decades, while simultaneously giving space to the assumption that, because not all national associations existing in the early 1950s joined AFC, the relations with the associations have been historically tarnished. In fact, the majority of the founding members came from east and southeast Asia, while there was no country from west Asia involved. It was in this context that the associations from east and southeast Asia came to feel closer-related to AFC, with the result that Malaysia played a strong role in administrative terms through hosting the AFC offices and seeing several of its citizens representing AFC as its President. These circumstances have also led to the emergence of power blocs, political power games in the run-up to elections, and critics stating that AFC should be divided. On the other hand, AFC's philosophy of advocating a pan-Asian outlook has been substantiated by its member associations on the global stage, namely with regard to increasing the quota of Asian delegates on FIFA's boards or the World Cup places for Asian teams.

In this sense, AFC can be viewed as an interface between local and global driving forces. Being part of the global infrastructure of football, it is required to accord with the FIFA Statutes and its regulations, while also representing and defending the interests of its members. Conflicts have occurred and continue, for instance in relation to the status of member associations, while collaboration is sought in terms of coordinating the international match calendar, development programmes and fighting match fixing. On the one hand AFC seeks to commit to FIFA's objective to harmonise the institutional framework, and on the other hand it aims to protect the status of its members under consideration of the particular circumstances in the country, and to consult with FIFA for example in cases of governmental interference in order to find a solution, or with regard to changing the Laws of the Game as in the case of the hijab rule.

AFC deviates from other confederations in its handling of the number of foreign players allowed in club sides and in its multi-tier competition system for national sides. It competes with other confederations in terms of promoting its competitions and making them more successful than others, as well as with regard to gaining more political influence at the global level, while seeking collaboration in the technical sector or creating mutual competitions. In fact, AFC refers to UEFA as a role model and at times takes a rather European perspective, defining European football as a benchmark and claiming that the exploitation of the Asian market through European clubs and leagues, their popularity and media presence may positively affect football in Asia. On the other hand, AFC stresses the need to develop a distinct Asian football culture and, for example, intervened when Manchester United wanted to tour Malaysia during the Asian Cup.

AFC's vision that the future of football lies in Asia is supported by its self-understanding as a developer of the game. For this purpose it promotes further professionalisation, commercialisation and mediatisation. Against the backdrop of having enhanced its own competitions and activities, not only structurally but also through partnerships with WSG and other commercial stakeholders, AFC is strongly interested in closing the gaps amongst member associations and initiating respective programmes and policies. In this sense, the confederation sets standards and criteria, and opts for harmonising league and club structures. Its Vision Asia development programme and the establishment of a licensing system for coaches and referees constitute sport development regimes which impose standardised requirements upon the member associations. These measures, however, have resulted in tensions given that especially the big leagues and member associations perceive AFC's initiatives at times as intervention and therefore want to maintain their autonomy. The smaller member associations in turn are rather receptive, while occasionally not committing fully. The fact that all these processes encompass transnational and global challenges – such as how to host sport events, how to tackle match fixing or how to promote good governance – requires acknowledging the plurality of actors involved and implementing measures on a level playing field.

The results therefore affirm the supposed role of AFC as 'glocal player' with a multipolar directionality in a spectrum between promoting either a rather global or a rather local/regional (Asian) business, media and management-oriented outlook within its power relations and network constellations. The degree, however, does vary and may appear ambiguous depending on whether specific issues concern extra- or intra-Asian relations, as well as on the political, economic, social, cultural and historical context. Notions developed in the theoretical discourse on globalisation – such as the historical phase models, glocalisation, and the differentiation between homogenisation and heterogenisation – thus offer useful anchor points for analysing the roles of international sport governing bodies.

The role of AFC as a 'corporate actor'

In the 1970s, political science experienced the so-called 'institutionalist turn' which led to a reconsideration of the significance of institutions with regard to analyses of agency and structure, decision-making and deliberation, policy programmes and outputs. Neo-institutionalists have referred to the importance of examining individual and complex action units within institutional settings and environments. Coleman (1974), for instance, has pointed out how international organisations act independently through institutionalised frameworks, and he drew attention to the constellations between the interests of corporate and individual actors, the institutional structures and regulations, spheres of competence and influence. In this sense, policy studies have increasingly concerned themselves with the utilisation of polity as a possible variable for examining political procedures and policy formulations.

Differing in questions on how to view the relationship between institutions and behaviour, as well as how institutions emerge and change, three streams can be identified in this theoretical spectrum, namely rational choice institutionalism, sociological institutionalism and historical institutionalism. The latter compensates the others' neglect of the temporal dimension and emphasises the historicity of institutions and political processes. It sets focus on how institutional arrangements can be defined by continuities and path dependencies. In contrast to rational choice and sociological institutionalism, which view the relation between actors and institutions as either determined by actors or institutions respectively, historical institutionalism considers the time factor to be an important variable. Assuming that institutions are produced through singular historical constellations, including elements of continuity and dynamics, path dependencies and policy legacies are regarded as crucial for institutionalisation processes. In this sense, institutions are understood as formal or informal norms or conventions which are mostly associated with an organisational capacity. Institutions, however, are not considered the only variable for explaining political processes; socio-economic and ideological aspects are considered as important.

Another stream, which is relevant for this study, is actor-oriented institutionalism, an integrative research heuristics developed by Mayntz and Scharpf (1995) during the 1990s. Accordingly, institutions are understood as both dependent and independent variables without a determining impact. Instead, institutional factors are regarded as shaping action contexts. Corporate actors, which interact in these contexts, are defined as organisations capable of acting in a sector of society, which is characterised by a high degree of self-organisation. Individual actions on the micro-level, however, are considered as important, given that corporate actors' strategies may depend on actions by individual members or representatives. The approach follows a multi-level perspective according to which an institutional framework shapes the actions of organisations, while the latter provide an institutional framework for the actions of their members. Institutions are therefore reduced to aspects of regulation and rules which can be intentionally designed and modified, while not determining but rather shaping and restricting actions. The two-fold perspective on actors and institutions allows overcoming the dichotomy between structure and agency, with observable actions functioning as proximate cause and institutional frameworks as remote cause. Institutions, however, can be viewed as programming occasions for interactions that follow a specific agendas or decision rules. Since institutional frameworks are seen as relevant with regard to shaping actors' orientations and aspects of situations in which actors operate, social entities such as organisations are regarded as suitable for being analysed institutionally, namely in relation to their incorporated rules and capability of acting as corporate actors. For this purpose, corporate actors are defined as formally organised bodies of persons which are capable of acting due to centralised resources that can be made use of upon hierarchic or majoritarian decisions.

Nonetheless, corporate actors consist of individual actors and it is thus important to examine the actions and orientations of individuals acting within and for a corporate actor. The orientations that actors have are viewed as shaped institutionally, as well as through the constellations that actors find themselves in. The strategic objectives, meanwhile, are not regarded as purely rationally driven, but rather by the context-free attributes of actors deriving from socialisation and general experience. The situations that actors operate in in turn generate a set of possible options which are not only influenced by the institutional framework but also by non-institutional factors, such as actually available resources.

With regard to sport politics and policies, Gammelsaeter and Senaux (2011: 3ff.) have drawn upon institutional organisation theory, concluding that actors (individuals, groups or organisations) are exposed to and shaped by institutional contexts. However, the actors are viewed as having capability of adapting to or modifying the norms and institutions. Decisions and outcomes are thus understood as products of interplays between agency and institutional structures. With regard to football, it is presumed that actors are associated with and dependent upon contradicting institutional environments, such as the civil sector, the market, the state and the sport sector.

The theoretical insights regarding the significance of institutions in relation to assessing the (inter-)actions by and between individual actors and corporate action units in specific constellations, as well as situations and decision-making processes resulting in policy outputs, are considered useful for analysing AFC in Asian and global football politics. While historical institutionalism provides grounds for looking at the historical developments, changes and continuities of and within AFC in the broader context of the history of football in Asia, actor-oriented institutionalism allows explaining and reconstructing interactions between actors in certain situations and constellations within an institutional framework. The definition of corporate actors, meanwhile, is useful with regard to analysing actions of and within AFC, as well as its relations with member associations and other actors. Analysing AFC institutionally therefore implies focusing on how institutional structures shape or restrict the actions of AFC, its members and its partners.

The analysed data reveals that AFC's raison d'être grounds on the planning, organisation and implementation of its competitions and tournaments, as well as friendly matches. This essential area of activity is not only restricted to the AFC's obligation and right to regulate and monitor the regulations of and compliance with the competition format, but also includes controlling the bidding process and election of a host country, as well as the proper organisational conduct by the local actors. In fact, from a historical perspective, AFC is founded on the idea of bringing together separate national associations, regionalising mutual interests and centralising resources for the sake of developing not only a unified voice for Asian football, but primarily to create an institutionalised frame (norms and rules, i.e. AFC's legislation and jurisdiction) with an organisational capacity that enables concerted efforts regarding implementation of tournaments and enforcement of respective regulations. The formation of AFC, however, was very much due to the initiative of individual actors with a pan-Asian outlook.

In principle AFC is not only a sum of its constituents, but the collectively authorised transfer of competences makes it an organisation capable of acting independently with regard to certain defined aspects. While membership in AFC is optional, it is necessary to join in order to participate in the established system of world football and the respective competitions. In fact, the historical development shows clearly how AFC has turned into the undisputable governing body of football in Asia due to a significant increase of members throughout the last decades. Plans for putting alternative structures into place have not materialised before today. However, this does not mean that member associations' powers vis-à-vis AFC have become obsolete. On the contrary, the member associations are not only represented in AFC's bodies and participate in the general assembly, the AFC Congress where they can initiate changes and modify the AFC's institutional framework, but according to the AFC Statutes they have the power to actually dissolve the confederation.

Being incorporated into the global governing system of football as defined by FIFA puts AFC in a global institutional framework, one which shapes its actions as well as the institutional structure at the regional level, which again shapes the actions of AFC's member associations and individual representatives. It is noteworthy that FIFA was rather sceptical when the confederations emerged in the mid-twentieth century and has ever since been determined to maintain a degree of control over them, thereby proving that taking the temporal dimension into account is necessary to understand current constellations, continuities and institutional dynamics. As a matter of fact, FIFA's aim to harmonise Statutes and regulations requires AFC to adapt and accord. FIFA does supervise, for instance, the AFC elections and Congresses, while being in the position to sanction the confederation. Nonetheless, AFC is an important actor in this relationship for it can initiate changes in the global institutional framework and does conflict with FIFA on selected matters. Examples include AFC's initiative to change the Laws of the Game with respect to wearing the hijab, and the fact that AFC's Statutes differ from FIFA's with regard to the term of the presidency as well as the quota of female representatives in the Executive Committee.

Regarding the decision-making processes within AFC, a mixture of path dependencies, continuities, changes and dynamics can be identified. While the institutional framework, in particular the AFC Statutes, has experienced extensive modifications throughout the last ten to 15 years, thereby changing the formal context in which procedures and policy formulations ought to evolve, the way of how decisions have de facto been prepared and taken has been constantly subject to the way individual actors have interpreted their posts, especially the Presidents and General Secretaries. In principle the AFC Congress is the supreme and legislative body, where all member associations convene in order to decide on the Statutes and membership affiliations, or elect representatives. All member associations have one vote and have the right to make proposals for the agenda of the Congress. However, it is in this context where again the tension between AFC as a unified voice and as a platform for power politics manifests itself. The Congress has been a site for political power games and for expressing the ambitions of individuals. Prominent examples include the conflicts on membership issues of and between countries such as China, Taiwan, Israel, Malaysia and Australia; the conflict with regard to changing the AFC headquarters; the conflict between Bin Hammam and Al Khalifa in the context of running for the seat on the FIFA Executive Committee; and the conflict between AFC and FAM with regard to hosting a match against Manchester United during the Asian Cup. However, the formal procedures are usually carried out rather smoothly and decisions are often unanimously adopted, while the informal agenda setting and exertion of influence takes place behind the scenes and open conflicts mostly arise in the run-up to the Congress or emerge afterwards.

While the executive body of AFC, the Executive Committee, is responsible for all matters not covered by the Congress or other bodies, it plays de facto a strong role in influencing the legislative outputs. In fact, the President, who is defined as the legal representative of AFC, can have a crucial stake in preparing and influencing the decisions of the Congress and the Executive Committee.

Having the deciding vote in the Executive Committee, he can interpret his role as an influential representative, including with regard to discussing items on its agenda. Moreover, his actions may have a considerable impact on the working procedures in the general secretariat. Therefore it is indeed important to assess the relation between actors and institutions, the significance of individuals at the political level, as well as the impact of the diversification and emergence of new actors on the aspirations and value system of AFC.

The results substantiate the supposed role of AFC as a 'corporate actor' that is capable of acting as an organisation within a dynamic institutional framework while simultaneously consisting of individual actors and engaging in relations with other actors, thereby shaping complex constellations all of which have to be understood in a historical perspective. The institutional framework does shape the actions of AFC, while the latter belongs to the institutional framework that influences the actions and decisions of its individual representatives and members, who again can initiate institutional changes or modifications. Ideas developed in the neo-institutionalist theoretical spectrum – such as the acknowledgement that institutions matter while focusing on actors, how both shape each other and create continuities and dynamics, as well as the definition of corporate actors – thus offer useful links for analysing the roles of international sport governing bodies.

The role of AFC as a 'skilful negotiator'

Responding to the emergence of new forms of governing – namely shifting away from hierarchic, state-oriented steering towards non-hierarchic forms of coordination between governmental and non-governmental actors across all levels – political scientists have increasingly paid attention to what has been labelled 'governance'. While the term lacks a unitary definition and has been understood differently depending on the scientific perspective employed, political science first related to the phenomenon in the context of international relations theories in order to assess power relations and structures that lack the competence to enforce binding decisions. Recognising these new forms of steering and coordination within and beyond national boundaries, also policy research studies has shown more interest in examining parallel and complementary forms of collective regulation, including self-regulation of private actors, co-regulation between public and private actors, and authoritative regulation though governmental actors.

Governance in this sense refers to mechanisms of regulation which are characterised by different forms of social interaction. These can include, for example, reciprocal adjustments in the market or independent actions, agreements in negotiations, or consent seeking in processes of communitisation. The translocations that occur in this context can be described as 'moving down', 'moving up' and 'moving out', i.e. the location of tasks and competences to subordinate structures, the shifting of competences to the supranational level, and the successive deregulation of governmental competences. Generally, governance is viewed as describing aspects of solving collective and complex societal problems. Therefore, differentiating between the three dimensions of polity, politics and policy has been considered useful by Benz in order to delineate governance from government perspectives. Accordingly, the governance perspective regards the state, the market and civil society as complementing each other in coordinating processes. The institutional structures provide the platform for conflicts, negotiations and adjustments between the different types of actors. In this context, the role of corporate actors indicates the challenges actors confront when transferring resources and competences to an organisation. For the purpose of identifying the complexity of structural and institutional arrangements, and their significance for examining the autonomy of members or representatives as well as the respective goal achievement probability, the definition of corporate actors offers a useful analytical approach, in particular with regard to INGOs.

Two specific types of governance have been referred to as multi-level and global governance. While the former relates to steering and coordination processes across various levels, differentiating between institutions as a set of rules and actors acting within these structures and presuming that competences are distributed across various levels, global governance offers an analytical perspective for conceptualising sector-specific forms of coordination through international organisations such as the WTO or WADA. Overarching all types of governance, however, are four core elements as defined by Benz (2004: 25ff.). These are: first, that governance signifies steering and coordination aimed at managing interdependencies between actors; second, that these processes are based on institutional control systems directing the actions of actors; third, that governance involves interaction patterns and modes of collective action that emerge in the context of institutions; and fourth, that the processes of steering and coordination, as well as interaction patterns covered by the term governance, usually transcend organisational boundaries and in particular the boundaries of state and society. Assuming that political processes are formed by cooperation between governmental and non-governmental actors, Benz recommends employing an institutionalist, actor-oriented and historical perspective that allows assessing dynamics of institutional contexts and how actors interact, adjust and collaborate in these structures.

With regard to sport studies, governance is considered a useful concept due to the traditionally non-commercial and non-governmental function of sport in society and the fact that the way it is organised is characterised by autonomous self-regulation through federations which are in principle separate from governmental influence, albeit receiving public funding and support. Defined as belonging to the third sector, sport can be viewed as a prototype of governance. Meanwhile, the increasing commercialisation and professionalisation of the sport sector has led to a diversification of actors and transgression of systemic boundaries, resulting in a structural hybridity characterised by new forms of steering and coordination. In this sense, the emergence of global sport policy networks relating to doping, racism and integrity issues constitute notable entry points. Postulating an acceptance for considering both structures and institutions, as well as beliefs and ideas, as anchor points for further studies, Grix (2010: 169–170) sums up that governance is a useful tool for examining the complexity of political processes.

Based on these assumptions, the analytical usefulness of governance for this study is viable in order to understand non-hierarchical forms of steering and coordination between governmental, economic and civil society actors framed by a variety of different institutional control systems and modes of interaction. In fact, the infrastructure of football implies a complex web of interactions between various actors such as federations, businesses, marketing agencies and governmental actors. Solving problems such as doping or match fixing requires coordination between different types of actors and appreciating the multi-layered diversity of interactions taking place between them.

Looking at with whom AFC interacts and how it does so, the results demonstrate that the developments of the last ten to 20 years are indeed characterised by the emergence of a diversity of actors and thus a web of different types of interaction, interests and spheres of influence. However, the modes of coordination and steering, as well as whether these are vertical or horizontal, relates to the specific constellation of actors and the item it is concerned with. From a historical perspective and looking through a pure sport sector-related lens, AFC is basically part of the self-regulated global football governing system, including clear definitions of formal decision-making procedures and divisions of power. In this sense, FIFA is at the top of the pyramid followed by the confederations and the member associations. AFC has to accord with the FIFA Statutes and the member associations have to fulfil AFC's requirements in order to essentially guarantee a smooth running of the game. This system has its own jurisdiction and in principle it forbids any interference through governmental actors. Nonetheless, even within this system AFC can function as an interface between FIFA and the member associations through modes of negotiation, agreement and mediation. The authority of the respectively higher organisation, though, remains undisputed, in particular with regard to disciplinary matters and sanctions. When it comes to implementing measures for instance in the technical area, though, FIFA, AFC and the member associations aim to coordinate their efforts in order to make the outcomes as effective as possible. The same applies to the organisation of competitions and the definition of the international match calendar.

Globalisation, commercialisation, professionalisation and mediatisation, meanwhile, have affected the sport sector. New challenges have resulted in an increase of cross-sectional constellations, collaborations and conflicts, in the course of which different institutional control systems and modes of interaction have become relevant. For instance, AFC engages increasingly with governmental actors not only in cases of interference in the autonomous sport sector, but also with regard to hosting tournaments, match fixing and corruption, doping abuse, and changing legislation for professionalisation measures. In turn, governments and civil society actors have progressively shown more interest in making sport governing bodies more accountable with respect to matters of good governance and integrity.

Meanwhile, the development of the game, the optimisation of grassroots and youth football structures, as well as the utilisation of football for social development purposes have led AFC to interact, as one amongst many, with a variety of actors including governments, other confederations, FIFA, member associations, civil society organisations and commercial enterprises. The Vision Asia development programme constitutes a good example for how important it is for AFC to act as a 'skilful negotiator' in order to ensure successful outcomes, for it is a consultancy service aimed at developing a mutual strategy and implementing a plan through entering a dialogue with member associations, governments and other actors at various levels. Another new constellation has developed due to the founding of regional federations. These are formally not recognised by FIFA or AFC, but they organise their own competitions, are represented by people holding AFC posts and gain more political influence. It will be interesting to see how AFC will interact with these federations in future.

Most notable is obviously the significance the market sector and therefore economic and commercial actors have attained throughout the last few decades. The progress of professionalisation in Asian football is strongly connected to whether and how successfully member associations, leagues and clubs have adapted to commercialisation processes. AFC has therefore made this aspect one of its major objectives. The fact that the confederation has entered deals with WSG and other commercial actors, generating large amounts of income, indicates this development. Due to the economic significance of the game, clubs, leagues and players have adopted new perspectives and demands. While Asia does not yet have a counterpart to the European Professional Football Leagues (EPFL) and European Club Association (ECA), FIFPro has its own division for Asia and Oceania. In fact, more and more countries see players' unions emerging. This development underlines the fact that AFC does have to expand its perspective and confront itself with new actors and requirements, for instance with regard to the contractual situation of players, the transfer system and the prevention of match fixing and bribery.

The results therefore assert the supposed role of AFC as a 'skilful negotiator' in the local, regional and global structure of football. This system, however, is not only characterised by a complex web of interrelations containing a high degree of diffusion of competence and authority amongst players, clubs, federations, broadcasters, national governments, fans, media, sponsors or agents. While AFC can be viewed as one element amongst many, with its structural and ideological sphere of influence varying and relying on functioning as a suitable coordinator and facilitator between all relevant actors, it has to be stated that, depending on the specific case, outputs are produced either in a rather vertical or horizontal mode of interaction, thereby indicating a tension between the traditional governing system and the emergence of new governance structures. Aspects identified in the discourse on governance – namely non-hierarchical forms of steering and coordination, cross-sectional constellations, a complex web of interactions between various actors and new modes of interaction – thus offer helpful links for analysing the roles of international sport governing bodies, while the assumption that a new governance system has fully replaced vertical structures and hierarchies cannot be entirely upheld.

Conclusions and outlook

Corresponding with a trend in sport policy studies to expand along a pluralist understanding of political relations and processes involving actors and institutions from different societal sectors at the local, national and international levels, this study contributes to going beyond the idea of treating sport as an arena for external political actors by focusing on the political processes and outputs within the sport sector. Recognising the general significance of non-governmental actors in global politics in an era of deregulation as well as liberalisation, emphasis has been put on how politics and policies take place within, through and for sports. The formulation and enforcement of specific interests, as well as the production and supervision of binding decisions through non-governmental actors in and for the sport sector, indicate that political aspects are part of the autonomous and self-regulated system. However, this does not exclude taking the interdependencies and the roles of governmental and other actors into account when analysing the diverse and multi-layered infrastructure of sport politics. For this purpose, the consideration of the three analytical categories polity, politics and policy was viewed important in order to contextualise the case study of AFC.

Employing a wide analytical lens that acknowledges the emergence of new forms of politics including new actors and structures, the study has applied definitions of INGOs and transnational organisations to AFC. While the exact differences are subject to debate, it is presumed that AFC can be considered a non-governmental organisation with the legal status of a non-profit entity, a constituency of 47 member associations and a political responsibility that ranges beyond national borders. However, it deviates from other INGOs in as much as its actions are mostly restricted to the Asian continent. Furthermore, the fact that football and in particular major tournaments have experienced a considerable commercialisation sets football governing bodies apart from other INGOs, thereby constituting an exceptional status with regard to scientific analyses.

The theory-led role of AFC as a 'glocal player' has shed light on the historical context in which AFC and football in Asia have emerged, its political location in an Asian and global context, its power and outreach in terms of harmonisation and regulation, the commercialisation of football in Asia and the way AFC engages in it, as well as how AFC is affected by feedback mechanisms and challenged by other driving forces. The role of AFC as a 'corporate actor' has illustrated the significance of institutional structures that the confederation acts in, how these shape or restrict the actions of the organisation, its members and representatives, how procedures and decision-making dynamics and continuities occur, as well as how certain actions may affect the institutional framework. The role as a 'skilful negotiator', meanwhile, has drawn attention to AFC's interactions with other actors in a horizontalised sport system, to the respective mechanisms of coordination, adjustment and negotiation, values and ideas, and the problem-solving procedures concerning specific issues in a multi-layered infrastructure characterised by dynamic institutional contexts.

The discussion of the roles and the theoretical perspectives they were based upon, under consideration of the empirical data, has indicated that AFC's actions unfold in various settings, constellations and situations. The multi-layered infrastructure, within which AFC's location ranges from finding itself in clearly defined vertical hierarchies to being part of a complex web of equal partners, makes it impossible to apply only one role to AFC. In fact, by its very nature it has always been at the intersection between at least FIFA and the member associations. And the diversification of actors and challenges in recent years – due to globalisation, commercialisation, mediatisation and professionalisation processes – has increased the variety of constellations AFC is involved in. This diversity of interfaces requires AFC to not only play a single role but to draw upon a set of roles. The three identified roles in this study are considered a set which describes and explains the functions and actions of the confederation certainly not entirely, but as comprehensively as possible. As has been demonstrated extensively, the theoretical perspectives all offer useful notions, provide sound assessments of the circumstances and encompass traits that are interconnected. Therefore, all roles are interconnected with each other, while it may not be universally predictable under which conditions AFC performs which role. However, the study has offered a coherent frame, and it drew upon indicators which affirm the supposed roles and make them applicable to the analysed cases and contexts.

The obtained roles correspond with how AFC understands itself, how it is viewed by others and how it is influenced by institutionalised structures. While AFC regards itself as a pan-Asian, proactive, future-oriented developer of the game, it may be perceived by some of its member associations as too authoritative and rather be expected to employ a more sensitive style. The conflict and reciprocity between top-down and bottom-up find its expression also in the important intersection between the local and the global, for example between FIFA, AFC and the member associations. Thus it can be concluded that AFC draws upon a set of roles which are shaped by ego and alter expectations, prescriptions, conflicts, adaptation, context-specific issues or institutionalisation processes.

Adding to the beneficial insights of globalisation, neo-institutionalism and governance, helpful insights with regard to studying sport governing bodies are provided by international relations theories and theories of integration, cooperation and regionalisation. International relations theories have been applied to studies on national and international sport policies, the governmental and non-governmental actors involved in the global infrastructure of sport and the extent of autonomy sport organisations possess. In fact, the importance of studying the relation between the autonomous sport sector and the governmental and public spheres has increased, given pressing themes such as the preservation of the integrity of sport or the sustainability of sport mega events. Meanwhile, theories of integration, cooperation and regionalisation provide fundamental points upon which to theorise the complex interrelations between all actors and levels. In particular, studies on the European Union can offer a useful apparatus. Presuming that for analytical purposes AFC could be equated with a regional political apparatus such as the EU, these studies and respective theories provide tools to examine the interplay between supranational and intergovernmental dimensions (FIFA, AFC and the member associations), the transfer of powers, the democratic legitimacy and accountability, questions on membership benefits and deficits, reactions and adjustments of members, the allocation of development funds, and aspects of identity politics.

To sum up, the added value of this study is threefold. First, it deals with a case which has never before been examined in as much detail, and it provides a newly collected and broad range of data. Second, it performs a systematic study on AFC from a political science perspective and delivers an analysis of its role. Third, it contributes to the development of sport policy studies as a sub-discipline through demonstrating that the three selected theoretical perspectives constitute a useful kit to work with. Other theories referred to, meanwhile, have additional potential, which it is recommended to be drawn upon more extensively in upcoming research, in particular when it comes to making use of insights produced through studies on the EU and other regional networks. In terms of the concept and analytic set-up, this work demonstrates how to draw upon political and social science as a toolkit for advancing sport policy research in sport science, while indicating the need to further interdisciplinary and international perspectives on sport politics and policy. As for the practical applicability, it not only serves as an assessment but also indicates how sport policy research can be utilised as a basis for policy advice.

The limitations of the study in turn derive from the fact that it is concerned with merely one case. It can therefore not claim comparative let alone universal validity. Nonetheless, it does offer an analysis of the usefulness of a theoretical spectrum to work with in relation to other cases. Moreover, the relatively broad perspective calls for more in-depth analyses of specific sub-cases, constellations and aspects, while the fact that AFC had not been studied in detail before required a basic and comprehensive research approach. The theoretical perspectives are very useful but they cannot assess the complexity of reality entirely. They do, however, contain a high degree of congruence and accuracy, and constitute a user-friendly framework and toolkit. Anyhow, more empirical studies are necessary, in particular with regard to the usefulness of governance concepts, for this study cannot affirm that a new governance system has fully replaced vertical structures and hierarchies. Considering the significance of individuals in the political processes, the analytical categories of polity, politics and policy could be extended to a fourth, namely politicians.

Finally, with regard to further specific topics for future research, it is recommended to conduct studies on specific policy outputs and outcomes, cross-sectional policy issues such as match fixing or mega sport events, selected bi- or multilateral relations with other confederations, regional federations or member associations, the harmonising effects of and reactions to AFC's development and professionalisation measures, the wider context of Asianism and identity, the role of women's football and the history of the ALFC, the time and period of transition following the presidency of Bin Hammam, as well as biographies of individuals or in-depth assessments of specific organs and bodies.

References

Amara, M., Henry, I., Liang, J. and Uchiumi, K. (2005). 'The governance of professional soccer: Five case studies – Algeria, China, England, France and Japan.' *European Journal of Sport Science*, 5(4), 189–206.

Benz, A. (2004). 'Governance: Modebegriff oder nützliches sozialwissenschaftliches Phänomen.' In A. Benz (ed.), *Governance: Regieren in komplexen Regelsystemen* (pp. 11–28). Wiesbaden: VS Verlag für Sozialwissenschaften.

Coleman, J. S. (1974). *Power and the structure of society* (1st edition), New York: Norton.

Gammelsæter, H. and Senaux, B. (2011). 'Perspectives on the governance of football across Europe.' In H. Gammelsæter and B. Senaux (eds), *Routledge research in sport, culture and society, Vol. 3: The organisation and governance of top football across Europe: An institutional perspective* (pp. 1–16). London: Routledge.

Giulianotti, R. and Robertson, R. (2007). 'Recovering the social: Globalization, football and transnationalism.' *Global Networks*, 2(7), 166–186.

Giulianotti, R. and Robertson, R. (2009). *Globalization and football* (1st edition). Los Angeles: SAGE.

Grix, J. (2010). 'The "governance debate" and the study of sport policy.' *International Journal of Sport Policy*, 2(2), 159–171.

Horne, J. and Manzenreiter, W. (2004). 'Football, cuture and globalisation: Why professional football has been going east.' In W. Manzenreiter and J. Horne (eds), *Football goes east: Business, culture and the people's game in China, Japan and South Korea* (pp. 1–18). London: Routledge.

Houlihan, B. (2009). 'Mechanisms of international influence on doemstic elite sport policy.' *International Journal of Sport Policy*, 1(1), 51–69.

Maguire, J. (1994). 'Sport, identity politics, and globalization: Diminishing contrasts and increasing varieties.' *Sociology of Sport Journal*, 11, 398–427.

Maguire, J. (1999). *Global sport: Identities, societies, civilisations*. Cambridge: Polity Press.

Maguire, J. (2006). 'Sport and globalization.' In J. Coakley and E. Dunning (eds), *Handbook of sports studies* (pp. 356–369). London: Sage Publications.

Mayntz, R. and Scharpf, F. (1995). 'Der Ansatz des akteurszentrierten Institutionalismus.' In R. Mayntz and F. Scharpf (eds), *Gesellschaftliche Selbstregelung und politische Steuerung* (pp. 39–72). Frankfurt am Main, New York: Campus Verlag.

Robertson, R. (1990). 'Mapping the global condition: Globalization as the central concept.' *Theory, Culture and Society*, 7(2–3), 15–30.

Index

Page numbers in *italics* denote tables.

ACL *see* AFC Champions League (ACL)
actor-oriented institutionalism 20–2, 23, 28, 176, 177, 180
Ad-hoc Committee for Indian Professional Football 101
ad-hoc committees, AFC 124, 134–5, 136
Administration System Department, AFC 139
AFAP *see* AFC Financial Assistance Programme (AFAP)
AFC *see* Asian Football Confederation (AFC)
AFC Champions League (ACL) 71, 82, 92, 127, 151; criteria for participating 56, 60, 103, 138; marketing 149, 152, 153
AFC Cup 103, 151, 153
AFC Diamond of Asia Award 105
AFC Financial Assistance Programme (AFAP) 91, 133, 136, 149, 161–3
AFC House, Kuala Lumpur 81–2, 105, 125
AFC Marketing Limited (AML) 71, 81, 106–7
AFDP *see* Asian Football Development Project (AFDP)
AFF *see* ASEAN Football Federation (AFF)
Afghanistan Football Federation *94*, *98*, 125, 126, 162
Africa *see* Confederation of African Football (CAF)
Afro-Asia competitions 91–2
Ahmad Shah, Sultan of Pahang 55, 81, 82, 89, 95, 105, 133
Ahmadinejad, Mahmoud 64, 65
AID27 Programme *see* AFC Financial Assistance Programme (AFAP)
AIFF *see* All India Football Federation (AIFF)
Al Khalifa, Salman Bin Ebrahim 126, 134
Al Sabah, Ahmad Falad 126–7
Al Sabah, Hasan 71
Al-Thawadi, Hassan 68–9
Alameh, Rahif 127
ALFC *see* Asian Ladies Football Confederation (ALFC)
Ali Bin Al Hussein, Prince 65, 90, 96, 104, 108, 129
All India Football Federation (AIFF) 58, 59, 60, *94*, *98*
All Manipur Football Association 61
All Nepal Football Association *94*, *98*
Allison, L. 17
Amara, M. 29–30, 31n2
AML *see* AFC Marketing Limited (AML)
ancient football traditions 38, 39
Anglo-Persian Oil Company 61
Anti-Doping Regulations, AFC 103–4, 140, 147–8
Appeals Committee, AFC 136–7
ASEAN Club Championship 95
ASEAN Football Federation (AFF) 93–5, *94*, 104, 129
Asian Club Championship 80, 82, 95
Asian Cup 101, 127, 151; Australia hosting of 100; China hosting of 51; FAO donation 164; Hong Kong hosting of 41; Iran in 63, 64; Lebanon hosting of 101, 106, 131; and Manchester United tour 102, 125; marketing 149; organisation of 153; Qatar hosting of 67, 149, 154; Saudi Arabia in 66
Asian Cup Department, AFC 153
Asian Cup Winners' Cup 81, 82

188 *Index*

Asian Football Confederation (AFC) 2–3; Anti-Doping Regulations 103–4, 140, 147–8; Code of Ethics 124, 127, 146–7; commercialisation 71, 81, 106–7, 182; Congress 38, 100, 121–9; as corporate actor 22–4, 175–9, 183; Disciplinary Code 124, 127, 140, 146, 147; educational programmes 136, 139, 155–8, 159, 160; Executive Committee 121, 122, 123, 124, 127, 128, 129–33, 134–6, 178–9; finance 126–7, 135, 140, 148–9; Financial Assistance Programme 91, 133, 136, 149, 161–3; founding of 15, 53; general secretariat 137–40; General Secretary 80, 121, 122, 130, 132, 133–4, 137–8; as glocal player 16–17, 171–5, 183; and governmental organisations 105–6, 145–6, 181; headquarters 81–2, 105, 125; history of 79–83, 173–4; as international non-governmental organisation 31n1; judicial bodies 136–7; jurisdiction and regulation 145–8; leagues, clubs and players 102–5; marketing 71, 81, 106–7, 123, 126–7, 149–50, 152; member associations 96–102, *98–9*, 121, 122, 145–6, 155–6, 162; non-governmental and civil society partners 107–9; President 121, 122, 123, 128, 129–30, 131, 133–4, 178–9; regional federations 93–6, *94*, 129, 131, 182; relations with FIFA 80, 85–91, 125, 145–6, 177–8, 181; relations with other confederations 91–3, 154, 174; representatives to FIFA 79, 82, 85, 86–7, 88–9, 90, 122, 123–4, 126, 128–9, 134; as skilful negotiator 29–30, 179–84; social responsibility activities 106, 108–9, 163–5; standing and ad-hoc committees 124, 128, 134–6, 146, 147–8; Statutes 121, 122, 123, 124, 127, 128–9, 137, 140, 146; technical matters 155–8; and women spectators in Iran 65; and World Cup (2002) 46, 47; *see also* competitions, AFC; Vision Asia development programme
Asian Football Development Project (AFDP) 108–9
Asian Games 38, 53, 58, 79
Asian Ladies Football Confederation (ALFC) 80, 87–8, 151
Asian Women's Football Championships 51, 80
Asian Youth Cup 65
Askew, D. 32n4

Aspire Academy, Qatar 67
Associacao de Futebol de Macau – China *94, 99*
Association Challenge Cup, Singapore 52
Association for the Promotion and Progress of Football, Iran 62
Association of Southeast Asian Nations *see* ASEAN Football Federation (AFF)
Audio-Visual Unit, AFC 140
Australia 95, 100, 104; *see also* Football Federation Australia (FFA)
Australia Soccer Association 100

Bahrain 145–6, 158, 160
Bahrain Football Association *94, 98*, 162
Bandyopadhyay, Kausik 4, 59–60, 61, 69
Bangladesh 160
Bangladesh Football Federation *94, 99*
barefooted players, India 58
baseball 40
Benz, A. 24, 26, 28, 180
betting 53, 70; *see also* match fixing
Bhutan Football Federation *94, 99*
Bin Hammam, Mohamed: accused of paying brides 83, 90, 134; at AFC Congress 38, 123, 124, 125, 126, 127, 128–9; and AFC Executive Committee 131–2; as AFC President 82–3, 126, 129, 133–4; on FIFA Executive Committee 82, 88–9, 123, 126, 134; and regional federations 95, 96; running for FIFA Presidency 82–3, 89–90, 95, 102, 134; and UEFA 92, 93
Black Whistle Affair, China 50
Blatter, Joseph "Sepp" 2, 88–90, 126, 127
Blazer, Chuck 90
Borja, David 72
Bosman ruling 15, 104
bribery scandals *see* corruption
British influence on Asian football 38, 39–40; China 49; India 39, 57–8; Iran 61–2; Japan 44; Malaysia 54; Singapore 52
Brunei 55; *see also* National Football Association of Brunei Darussalam
budgets, AFC 126–7, 135, 140, 148–9
Business Houses League, Singapore 52, 53

CAF *see* Confederation of African Football (CAF)
Calcutta League 57, 58
Cambodia *see* Football Federation of Cambodia
Careem, Nazvi 72

Caribbean *see* CONCACAF (Confederation of North, Central America and Caribbean Association Football)
Caribbean Football Union 83, 90
CAS *see* Court of Arbitration for Sport (CAS)
Castells, M. 17
Central America *see* CONCACAF (Confederation of North, Central America and Caribbean Association Football)
Centres of Sports Medicine Excellence 105
CFA *see* Chinese Football Association (CFA)
Challenge Cup 82
challenges for Asian football 69–73
Champions League *see* AFC Champions League (ACL)
Chelsea FC 108, 150
China 173; and AFC conflict with FIFA 87; ancient football tradition 38, 39; history of football 49–52; and Hong Kong 41; Vision Asia programme 160; women's football 2, 51
China Soong Ching Ling Foundation 108
Chinese Football Association (CFA) 49, 50–2, *94*, *99*, 108, 146
Chinese Professional Soccer League 50
Chinese Super League 50–1
Chinese Taipei Football Association 87, *94*, *98*
Chung Mong-Joon 42, 88, 89, 90, 129
Church, Michael 72–3
civil society 107–9, 181
Cleland, John M. 79
climate issues, Qatar World Cup 67–8
Close, P. 32n4
club licensing system 103
coach employment scheme 162
coaches: education 136, 155, 158, 159, 160; female 157–8, 162
Coaches Association, AFC 124
Coaching Committee, AFC 136
Code of Ethics, AFC 124, 127, 146–7
Code of Sports-related Arbitration 137
Coleman, J. S. 18, 21, 175
collective action, modes of 28, 180
commercialisation: AFC 71, 81, 106–7, 182; China 50, 51; India 60; Iran 64–5; Japan 173; South Korea 173
competitions, AFC 150–4; Afro-Asia 91–2; club 103; expenditure on 149, 153; history of 80, 81, 82; marketing 107, 149–50, 152; regional federations 95, 96; women's 51, 80, 88, 138, 149, 151; youth 80, 81, 149, 151, 162; *see also* AFC Champions League (ACL); Asian Cup
Competitions Committee, AFC 135–6
Competitions Department, AFC 138, 153
CONCACAF (Confederation of North, Central America and Caribbean Association Football) 92
Confederation of African Football (CAF) 4, 91–2
confederations 85, 91–3; CONCACAF 92; Confederation of African Football (CAF) 4, 91–2; CONMEBOL 86, 89, 92; Oceania Football Confederation (OFC) 92, 100; *see also* Asian Football Confederation (AFC); UEFA (Union of European Football Associations)
Congress, AFC 38, 100, 121–9
CONMEBOL (Confederación SudAmericana de Fútbol) 86, 89, 92
cooperation theories 185
coordination *see* steering and coordination
corporate actors 21, 176; AFC as 22–4, 175–9, 183
corporate ownership of clubs 42, 44, 53, 72
corruption 70, 126; accusations against Bin Hammam 83, 90, 134; AFC measures against 83, 106, 154, 157, 165; China 50, 51–2; Malaysia 56; and Qatar World Cup 68; Singapore 53; South Korea 43; *see also* match fixing
Court of Arbitration for Sport (CAS) 137
criminal networks 70
Cultural Revolution, China 49
Czempiel, Ernst Otto 25

Dalian Wanda 52
Darby, P. 4
Deng Xiaoping 49–50
Dentsu 47
development programmes, AFC 91, 101, 157–8; *see also* AFC Financial Assistance Programme (AFAP); Vision Asia development programme
Diamond of Asia Award, AFC 105
Dimeo, P. 4
diminishing contrasts and increasing varieties concept 14, 31n2, 171
Disciplinary Code, AFC 124, 127, 140, 146, 147

Index

Disciplinary Committee, AFC 136–7, 147–8
Dispute Resolution Chamber (DRC), FIFA 104
diversity of Asia 70, 173
Dodd, Moya 128
doping 103–4, 140, 147–8
Dorsey, James M. 69
DRC *see* Dispute Resolution Chamber (DRC), FIFA
Dream Asia programme 106, 108, 163–5
Durand Cup, India 57–8, 60
Dzung, Ms. 127

EAFF *see* East Asian Football Federation (EAFF)
East Asian Football Championship 95
East Asian Football Federation (EAFF) 93, *94*, 95, 129
East Bengal FC 58, 60
economic benefits of World Cup 47–8
Education Division, AFC 139
educational programmes, AFC 136, 139, 155–8, 159, 160
Eisenberg, C. 4, 11
Elias, N. 14
Emergency Committee, AFC 131
English Football Association (FA) 39, 107–8
Ethics Committee, AFC 124, 136–7
European football 72–3, 150, 153–4, 174; *see also* UEFA (Union of European Football Associations)
European Union 14–15, 29, 185
Event and Logistics Department, AFC 140
Executive Committee, AFC 121, 122, 123, 124, 127, 128, 129–33, 134–6, 178–9
expenditure, AFC 149, 153
Extraordinary Congresses, AFC 100, 121, 123, 124, 128

FA *see* English Football Association (FA)
FAM *see* Football Association of Malaysia (FAM)
FAO *see* Food and Agriculture Organisation (FAO), United Nations
FAS *see* Football Association of Singapore (FAS)
Federacao de Futebol Timor Loro-Se *94*, 99, 125, 126
Federation Cup, India 60
Federation Libanaise de Football Association *94*, 98
Fernando, Manilal 127, 129

Festivals of Football 125, 127, 139, 162, 163
FFIRI *see* Football Federation Islamic Republic of Iran (FFIRI)
FIFA 3, 4, 11; Anti-Doping Regulations 147; Bin Hammam's Presidency campaign 82–3, 89–90, 95, 102, 134; development programmes 161; Dispute Resolution Chamber 104; founding of 15; and Iran 64; and regional federations 93; relations with AFC 80, 85–91, 125, 145–6, 177–8, 181; representatives from AFC 79, 82, 85, 86–7, 88–9, 90, 122, 123–4, 126, 128–9, 134; rights of confederations 85; Women's World Cup Conference 157; and World Cup (2002) 47–8
FIFA Ethics Committee 83, 90
FIFA Executive Committee 79, 82, 85, 86–7, 88–9, 90, 122, 123–4, 126, 128–9, 134
FIFA World Youth Tournament 56
FIFPro 104, 182
finance, AFC 126–7, 135, 140, 148–9
Finance Department, AFC 140
Finance and Marketing Committee, AFC 135
Financial Assistance Programme *see* AFC Financial Assistance Programme (AFAP)
five-phase history of football model 15, 17, 172
five-phase sportisation model 14, 17, 171
Flath, Thomas 71–2
Fok, Henry 41
Fok, Timothy 131
Food and Agriculture Organisation (FAO), United Nations 106, 164
Football Association of Indonesia *94*, *98*, 100
Football Association of Malaysia (FAM) 53, 54–6, 88, *94*, *98*, 102, 125
Football Association of Maldives *94*, 99, 124
Football Association of Singapore (FAS) 53, *94*, *98*
Football Association of Thailand *94*, *98*, 100
Football Association of The Democratic People's Republic of Korea *94*, *98*
Football Association of Turkmenistan 99
Football Federation Australia (FFA) 99, 100, 124
Football Federation of Cambodia *94*, *98*, 162

Football Federation Islamic Republic of Iran (FFIRI) 62, 64, 65, *94*, *98*, 100
Football Federation of Kyrgyz Republic *99*
Football Federation of Sri Lanka *94*, *98*
foreign players, numbers allowed 103
Foreign Talent Scheme, Singapore 53
franchising system, South Korea 42
Futsal: competitions 149, 151; development programmes 158, 159; Iran 64, 65; regional 95, 96
Futsal Committee, AFC 136
Futsal Department, AFC 138–9
Futuro Coaching Programme 88, 91

Gaillard, William 93
gambling 53, 70; *see also* match fixing
Gammelsæter, H. 22, 176
general secretariat, AFC 137–40
General Secretaries Induction Programme 101
General Secretary, AFC 80, 121, 122, 130, 132, 133–4, 137–8; *see also* Samuel, Paul Mony; Soosay, Alex; Velappan, Peter
German Football Association 107
'germinal' phase in history of football 15
Giulianotti, R. 4, 13, 15–16, 17, 31–2n3, 39, 172
global governance 27–8, 180
globalisation 11–17, 31n2, 171–5
glocal player, AFC as 16–17, 171–5, 183
glocalisation 15, 16, 172
Goa Football Association 61
Goal Programme 91, 161
Goldblatt, D. 39
governance 22, 24–31, 165, 179–83
Government Services League, Singapore 52
governmental organisations, and AFC 105–6, 145–6, 181
grassroots football *see* youth and grassroots football
Grassroots and Youth Department, AFC 139, 160
Grix, J. 30–1, 181
Groll, M. 29
Große Hüttmann, M. 27
Guam Football Association *94*, *99*, 100, 162
Güldenpfennig, S. 10

Hall, P. 18
Hamzah Abu Samah 80–1
harmonisation 14, 16, 171, 172

Harnisch, S. 10–11
Havelange, João 3, 15, 47, 87
Hayatou, Issa 92
headscarf rule 65, 90, 104, 108–9, 133, 157
health care, players 105
heterogenisation 16, 171, 172
hijab rule 65, 90, 104, 108–9, 133, 157
Hill, Declan 70
Hindley, D. 29
historical institutionalism 19, 23, 28, 32n7, 175, 177, 180
history of football: Asia 38–40; China 49–52; five phases of 15, 17, 172; global 15–16; Hong Kong 40–1; India 38, 39, 57–61; Iran 61–5; Japan 39–40, 43–6; Malaysia 54–6; Saudi Arabia 65–7; Singapore 52–3; South Korea 41–3
HKFA *see* Hong Kong Football Association (HKFA)
Holt, M. 30
homogenisation 16, 171
homosexuality 165
Hong, Fan 38
Hong Kong 40–1, 79–80, 173
Hong Kong Football Association (HKFA) 40–1, *94*, *98*
Hong Kong Shield 40, 49
Horne, J. 4, 16, 172
Houlihan, B. 11, 14–15, 16, 171, 172
Human Resources Department, AFC 140
human rights issues, Qatar World Cup 68
human trafficking, Qatar World Cup 68
Hyderabad Football Association 57–8
Hyundai Research Institute 47

I-League 60, 61
IFA *see* Indian Football Association (IFA)
IFAB *see* International Football Association Board (IFAB)
IFRS *see* International Financial Reporting Standards (IFRS)
'incipient' phase in history of football 15
income, AFC 148–9
India 40, 173; Ad-hoc Committee for Indian Professional Football 101; All India Football Federation (AIFF) 58, 59, 60, *94*, *98*; history of football 38, 39, 57–61; players 104; Vision Asia programme 61, 160; women's football 58–9
Indian Football Association (IFA) 59, 61
Indian Football Association Shield 57, 58, 60

Index

Indonesia 40, 104; *see also* Football Association of Indonesia
Information Services Department, AFC 139
INGOs *see* international non-governmental organisations (INGOs)
institutional control systems 27, 28, 29, 180, 181
institutionalism 17–24, 28, 32n5, 32n7, 175–9, 180
integration theories 185
Internal Audit Committee, AFC 124, 135
Internal Audit Department, AFC 140
International Financial Reporting Standards (IFRS) 148
International Football Association Board (IFAB) 133, 157
international match calendar 152–3
international non-governmental organisations (INGOs) 11, 31n1
International Olympic Committee (IOC) 11, 50, 147
International Paralympic Committee (IPC) 147
international relations theories 184
International Sport and Leisure (ISL) 81, 106
IOC *see* International Olympic Committee (IOC)
IPC *see* International Paralympic Committee (IPC)
Iran 61–5, 158, 160, 161, 173; *see also* Football Federation Islamic Republic of Iran (FFIRI)
Iranian Pro League 64
Iraqi Football Association *94*, *98*
ISL *see* International Sport and Leisure (ISL)
Israel 63, 80, 87, 100

J-League 44–5
Japan: history of football 39–40, 43–6; and Korea 41; players 104; professionalisation 44–5, 154, 173; women's football 2, 45; World Cup (2002) 2, 40, 45, 46–8, 89
Japan Soccer League 44
Japanese Football Association (JFA) 44, 45–6, *94*, *98*
Japanese World Cup Organising Committee (JAWOC) 48
JFA *see* Japanese Football Association (JFA)
Johansson, Lennart 88

Jordan 108, 154
Jordan Football Association *94*, *99*
judicial bodies, AFC 136–7

K-League 42–3
Karimi, Ali 64
Käser, Helmut 88
Kawabuchi, Saburo 44
Kazakhstan 70, 100
kemari 39
Kenyon, Peter 108
Keohane, R. 13
KFA *see* Korea Football Association (KFA)
King's Cup, Saudi Arabia 66
Knodt, M. 27
Korea Football Association (KFA) 41, 42, 43, *94*, *98*, 100, 127
Korean Development Institute 47
Korean Super League 41–2
Kuwait Football Association *94*, *98*, 100, 125, 126, 128
Kyrgyzstan 70; *see also* Football Federation of Kyrgyz Republic

labour conditions, Qatar World Cup 68
Lahoud, Emile 101
Lai, Richard 126
Lao Football Federation *94*, *99*, 125, 126
Lebanon 101, 106, 131; Federation Libanaise de Football Association *94*, *98*
Lefkaritis, Marios 93
Legal Committee, AFC 124, 128, 136, 146
Legal Department, AFC 140
Lim Kee Siong 86
Lineker, Gary 44

Macau *see* Associacao de Futebol de Macau – China
Maguire, J. 13–14, 16, 17, 171, 172
Mah Bow Tan 53
Mahdavikia, Mehdi 64
Majumdar, Boria 4, 59–60
Makudi, Worawi 90, 95, 129
Malayan Chinese Olympic team 54
Malaysia 54–6, 104, 164, 173; AFC House, Kuala Lumpur 81–2, 105, 125; *see also* Football Association of Malaysia (FAM)
Malaysia Cup 52, 53, 54
Malaysia Super League (MSL) 56
Malaysian Accounting Board 148
Malaysian FA Cup 55

Maldives *see* Football Association of Maldives
Manchester United FC 102, 125
Manzenreiter, W. 4, 16, 45, 172
marketing: AFC 71, 81, 106–7, 123, 126–7, 149–50, 152; China 50, 51; India 60; Iran 64–5; Japan 44; World Cup (2002) 44
Marketing Department, AFC 140
match calendar, international 152–3
match fixing 70; AFC measures against 83, 100, 103, 106, 154, 157; China 50, 51–2, 83; Malaysia 56; Singapore 53; South Korea 43, 83
Mayntz, Renate 20–2, 23, 25, 26, 176
media 72–3, 139, 159
Medical Committee, AFC 124, 136
Meier, H. E. 20, 22, 23–4
member associations, AFC 96–102, *98–9*, 121, 122, 145–6, 155–6, 162
Member Associations Committee, AFC 146
mental health issues 165
Merdeka Tournament 53, 54, 55, 80
Midnight Football Programme, AFC 106, 163–4
migrant workers, Qatar World Cup 68
Mills, J. 4
missionary schools, Iran 61
Miyazawa Kiichi 46
modularised stadiums 68
Mohammedan Sporting 57–8
Mohun Bagan 38, 57, 58, 60
Mongolian Football Federation *94*, *99*
MSL *see* Malaysia Super League (MSL)
multi-level governance 27, 180
Murray, Bill 70
Myanmar 106, 154
Myanmar Football Federation *94*, *98*

Najib Tun Razak 105
naming of teams, South Korea 42
National Association for Physical Education, Iran 62
National Football Academy, Singapore 53
National Football Association of Brunei Darussalam *94*, *98*, 100
National Football League (NFL), India 57, 60
National Football League (NFL), Singapore 53
National Sports Commission (NSC), China 49
National Sports Council (NSC), Malaysia 55, 56

National Women's Sport Association, Iran 65
neo-institutionalism 17–24, 28, 32n5, 32n7, 175–9, 180
neo-mercantilism 31–2n3
Nepal *see* All Nepal Football Association
new public management 24–5
NFL *see* National Football League (NFL), India; National Football League (NFL), Singapore
non-governmental sector 107–9
North America *see* CONCACAF (Confederation of North, Central America and Caribbean Association Football)
Northern Mariana Islands Football Association *94*, 95, *99*
NSC *see* National Sports Commission (NSC), China; National Sports Council (NSC), Malaysia
Nye, J. 13

O'Brien, Seamus 106
Oceania Football Confederation (OFC) 92, 100
Okano Shun'ichiro 47
Olympic Games 41, 49, 51, 54, 55, 58, 62; *see also* International Olympic Committee (IOC)
Oman 160
Oman Football Association *94*, *99*
Oods Cup, Iran 63
Ordinary Congresses, AFC 121, 123–8, 129
organisational governance 26–7

Pakistan 160
Pakistan Football Federation *94*, *98*
Palestinian Football Association *94*, *99*
Pereira, Charles D. 88
Performance Programme, FIFA 161
PFA *see* Professional Footballers Australia (PFA)
Philippine Football Federation *94*, *98*
Philippines 109, 160
players: Asian international 2; improving conditions for 104–5, 165; numbers of foreign allowed 103; sanctions for doping abuse 147–8; unions 104, 182
Players' Status Committee, AFC 124
politics and sport 1, 9–11; globalisation 11–17, 31n2, 171–5; governance 22, 24–31, 165, 179–83; neo-institutionalism 17–24, 28, 32n5, 32n7, 175–9, 180; neo-mercantilism 31–2n3

President, AFC 121, 122, 123, 128, 129–30, 131, 133–4, 178–9; *see also* Ahmad Shah, Sultan of Pahang; Bin Hammam, Mohamed; Hamzah Abu Samah; Tunku Abdul Rahman; Zhang Jilong
President's Cup 82, 103, 151
Professional Coaching Diploma Course, AFC 92
Professional Football Development Taskforce 101, 103
Professional Football Project, AFC 138, 154, 161
Professional Footballers Australia (PFA) 104
Professional Leagues Project, AFC 82
professionalisation 15, 29, 40, 69–70, 71; of administrative structures 157; AFC schemes 82, 104, 105–6, 154, 182; China 50, 51; India 60, 61; Iran 64; Japan 44–5, 154, 173; Malaysia 55, 56; Saudi Arabia 66; South Korea 41, 154, 173; women's football 45, 51
Project Future 127, 155, 157

Qatar 67–9, 149, 154, 165
Qatar Football Association 67, *94, 98*
Qatar Stars League 67

rational choice institutionalism 18–19, 175
Reach Out To Asia (ROTA) 108, 165
Redeker, Robert 1
referees: education 139, 155, 156–7, 160; female 157–8, 162
Referees Association, AFC 124
Referees Committee, AFC 136
Referees Department, AFC 139
regimental clubs 52, 57–8
regional federations 93–6, *94*, 129, 131, 182
regionalisation theories 185
Reza Shah 62
Rhodes, R. A. W. 25
Rivelino, Roberto 66
Robertson, R. 4, 12–13, 15–16, 17, 31–2n3, 39, 171, 172
Ronge, V. 28–9
Rosenau, James N. 25, 28
ROTA *see* Reach Out To Asia (ROTA)
Rous, Sir Stanley 86
Rovers Cup, India 57, 60

S-League 53
SAFA *see* Singapore Amateur Football Association (SAFA)

SAFF *see* South Asian Football Federation (SAFF)
Saguto, Santino 67
Saito, Satoshi 150
Salahuddin, Kazi 96
Samuel, Paul Mony 2, 82, 95
Santosh Trophy, India 60
Saudi Arabia 65–7, 173
Saudi Arabian Football Association 66, *94, 98*, 100
Saudi League 66
SCAA *see* South China Athletic Association (SCAA)
Scharpf, Fritz W. 20–2, 23, 25, 176
Schmid, J. 10
Schneider, V. 27
school leagues 160
Schwab, Brendan 104
Scudamore, Richard 108
Security Committee, AFC 124
Selangor Amateur Football League 54
Selangor Football Association 54
Semi-Pro Football League, Malaysia 55
Senaux, B. 22, 176
Shanghai Athletic Club 49
Singapore 52–3, 55; *see also* Football Association of Singapore (FAS)
Singapore Amateur Football Association (SAFA) 52, 53, 54
Singapore Football League 52–3
skilful negotiator, AFC as 29–30, 179–84
social responsibility activities, AFC 106, 108–9, 163–5
Social Responsibility Committee, AFC 136
Social Responsibility Department, AFC 163, 165
socio-cultural benefits of World Cup 48
sociological institutionalism 19, 175
Soh Ghee Soon 53, 86
Soosay, Alex 83, 108, 133
South America *see* CONMEBOL (Confederación SudAmericana de Fútbol)
South Asian Football Federation Championship 58, 96
South Asian Football Federation (SAFF) 93, *94*, 95–6, 129
South China Athletic Association (SCAA) 49
South Korea: history of football 41–3; Korea Football Association (KFA) 41, 42, 43, *94, 98*, 100, 127; professionalisation 41, 154, 173;

women's football 43; World Cup (2002) 2, 40, 46–8, 89
sponsorship, AFC 107, 148–9, 150, 152
sportisation model, five-phase 14, 17, 171
sports medicine 105
Sri Lanka 160; *see also* Football Federation of Sri Lanka
standing committees, AFC 128, 134–6, 146, 147–8
Standing Orders of Congress, AFC 121
Statutes, AFC 121, 122, 123, 124, 127, 128–9, 137, 140, 146
steering and coordination 22, 24–7, 28, 180, 181
Steinbach, D. 29
Steuerungstheorie 25
Stoker, G. 25
'struggle-for-hegemony' phase in history of football 15, 172
Sugden, J. 4, 11, 17, 29
Super Cup 81, 82
Suzuki Cup 95
Suzuki Tokuaki 93, 153–4
Syrian Football Association *94*, *98*

Taiwan 80, 87; *see also* Chinese Taipei Football Association
Tajikistan 70
Tajikistan Football Federation *99*
'take-off' phase in history of football 15, 39–40
Tapa, Ganesh 96, 127
Tashima, Kohzo 95, 126
taxation, AFC 149
Taylor, R. 18
Technical and Vision Asia Committee, AFC 136
temperature issues, Qatar World Cup 67–8
Tengku Mahkota 56
Thailand 105, 156, 160, 161; *see also* Football Association of Thailand
Tiger Cup 53, 95
Timor Leste *see* Federacao de Futebol Timor Loro-Se
Tokarski, W. 29
Tomlinson, A. 4, 11, 17
transnationalism 13
transparency 90
Triple Crown, India 57
Tunku Abdul Rahman 54–5, 80, 105
Turkmenistan 70; *see also* Football Association of Turkmenistan
'Two Chinas' policy 50

UEFA (Union of European Football Associations) 4; founding of 15; Israel joins 87; relations with AFC 92–3, 154, 174; representatives to FIFA 86; support for Blatter 90; and World Cup (2002) 47, 89
'uncertainty' phase in history of football 15, 172
United Arab Emirates 131
United Arab Emirates Football Association *94*, *99*
United Nations Food and Agriculture Organisation (FAO) 106, 164
United States 68
Uzbekistan 70, 160
Uzbekistan Football Federation *99*

Valcke, Jerome 126
Velappan, Peter 69–70, 71, 80–1, 87, 93, 95, 106, 125, 132, 133–4
Vice-Presidents, AFC 122, 129
Vietnam 160
Vietnam Football Federation *94*, *98*
Vision Asia Department, AFC 139, 160
Vision Asia development programme 3, 16, 82, 159–61, 182; and Bin Hammam 82, 134; and Chelsea FC 108; expenditure on 149; and FIFA 91; and governmental organisations 106; in India 61, 160; sponsorship of 148, 149; Technical and Vision Asia Committee 136; and UEFA 92; youth and grassroots football 156, 159, 160

WAFF *see* West Asian Football Federation (WAFF)
Weber, Max 30
Website Department, AFC 139
West Asian Football Federation Championship 96
West Asian Football Federation (WAFF) 93, *94*, 96, 129
women spectators, Iran 64, 65
Women's Asian Cup 138, 151
Women's Committee, AFC 136
Women's Department, AFC 138, 157
women's football 2; Asian Ladies Football Confederation (ALFC) 80, 87–8, 151; China 2, 51; competitions 51, 80, 88, 138, 149, 151; development programmes 157–8, 159, 162; female coaches and referees 157–8, 162; hijab rule 65, 90, 104, 108–9, 133, 157; India 58–9; Iran 64–5; Japan 2, 45;

women's football *continued*
  professionalisation 45, 51; regional 95, 96; South Korea 43
Women's Football Federation, South Korea 43
Women's Soccer Federation, India 59
Women's World Cup 43, 51, 88
World Anti-Doping Code 103, 147
World Cup 15; Asian qualifiers 88, 149, 151; China in 50; India in 58; Iran in 64; Italy (1990) 40; Japan in 44; Qatar (2022) 67–9, 165; Saudi Arabia in 66; South Korea and Japan (2002) 2, 40, 45, 46–8, 89; South Korea in 41, 42; *see also* Women's World Cup
World Sport Group (WSG) 71, 81, 107, 126–7, 148, 150, 152

Woyke, W. 31n1
WSG *see* World Sport Group (WSG)

Yemen Football Association *94*, *98*
youth and grassroots football: AFC policies 155–6; China 50, 52; competitions 80, 81, 149, 151, 162; Grassroots and Youth Department 139, 160; India 61; Malaysia 55; Qatar 67; school leagues 160; and Vision Asia programme 156, 159, 160

Zhang Jilong 83, 90, 95, 96, 129, 132
Zico 44
Zwanziger, Theo 68